What If
the Earth
Had
Two Moons?

WHAT IF THE EARTH HAD

TWO MOONS?

And Nine Other

Thought-Provoking

Speculations on

the Solar System

Neil F. Comins

St. Martin's Press ✖ New York

www.stmartins.com

Library of Congress Cataloging-in-Publication Data

Comins, Neil F., 1951–
    What if the Earth had two moons? : and nine other thought-provoking speculations on the solar system / Neil F. Comins.—1st ed.
        p. cm.
    Includes bibliographical references.
    ISBN 978-0-312-59892-1
    1. Earth—Miscellanea.   2. Solar system—Miscellanea.   I. Title.
    QB638.C66 2010
    525—dc22   2009040247

First Edition: April 2010

10   9   8   7   6   5   4   3   2   1

Dedicated with love to my wife,
SUZANNE, who is still my girlfriend after all these years.

To my son JAMES, who asked lots of "What if" questions;
to my son JOSHUA, who also asks "Why"; and
to our dear friends WALTER, SANDRA, and ELIAS LYND.

I hope that the scenarios in this book give us
glimpses into the environments of other planets in the universe
and the conditions that may exist on those worlds.

Contents

Acknowledgments

My warmest thanks to my editor at St. Martin's Press, Phil Revzin, to W.H. Freeman and Co. for allowing me to use art from some of my other books in this one, to Alan Boss for his insights into giant planet formation, to David Sturm for discussions about several of these scenarios, and to my wife, Sue, for her love and patience as I wrote this book.

What If
the Earth
Had
Two Moons?

Introduction

Have you ever heard a child ask a question beginning, "What if . . ."? Fortunately for me, my older son James seemed to do that every few minutes when he was young. During that epoch my colleague, Dave Batuski, observed to me that we astronomers are always looking at the cosmos from the same old perspectives. "Well," I thought, "how does one look at it differently after decades of running the same scripts?" James's questions provided the answer. Ask "What if" questions and explore the consequences. The first such question that popped into my mind was: "What if the Moon didn't exist?" Within minutes I realized that the consequences of such a change for the Earth and life on it are staggering. I began to research and write about this and other "What if?" questions.

In 1991, *Astronomy* magazine published my article entitled "What If . . . Earth Had No Moon?," and followed this with several other "What if?" scenarios, each exploring a new version of Earth and showing how it differs from our world. These and other scenarios were published in the 1993 book, *What If the Moon Didn't Exist?*, which was described as a new genre of science writing by *Washington Post* writer Curt Suplee. *What If the Moon Didn't Exist?* has since been used in radio and television shows; it has been made into planetarium shows; it has been taught as college courses; it was chosen as the theme of Mitsubishi's pavilion at the 2005 World Expo in Japan (seen by three million people); and it was made into a permanent show at a resort in southern Japan.

The "What if?" concept has now morphed into "makeovers." Many people find it fascinating to see how a place or person can be transformed "with a little help from their friends." What would this house look like if it were made over by professional designers? What would this person look like if his or her weight were changed or wardrobe redone? Indeed, "What if?" questions have the power to change lives and move worlds. In fact, most of us ask these questions in our everyday lives: "What if I take that job?" and use them to explore the consequences: "Then I will be able to pay the mortgage." "What if I take the cruise?" "Then maybe I will meet Miss Right." "What if I hadn't tried that climb?" "Then I wouldn't have broken my ankle." Asking and answering such questions often gives us valuable insights into our decision-making processes and the wisdom or folly thereof.

The power of "What if?" questions doesn't stop there. Historians ask them in order to explore how society might have evolved under different circumstances. "What if the South had won the U.S. Civil War?" "What if Napoleon had been victorious at Waterloo?" "What if Al Gore had won the White House in 2000?" The complexities of personal, interpersonal, and societal behaviors prevent us from being absolutely sure of the conclusions we draw from such personal or historical "What if?" questions. Indeed, in the TV makeover world, altered contestants have seldom undergone permanent changes.

Scientists often fare better in understanding and exploring alternative situations. We ask questions for which experiments, calculations, and observations let us judge their consequences. "What if I add three percent more potassium to this solution?" "Then the reaction rate increases by fifteen percent." "What if the Sun had started evolving with ten percent more matter than it actually did?" "Then it would be brighter by forty percent." "What if I alter this segment of DNA?" "Then the organism would be blind." Exploring these alternative scenarios often provides immediate insights, as well as ideas for alternative experiments and calculations.

Interest in alternative versions of Earth leads into the intriguing realm of whether there are other life-supporting planets in our Milky Way Galaxy. This search begins right in our backyard, as astronomers have virtually incontrovertible evidence that liquid water, essential for life as we know it, existed in great quantities on Mars. Could there still be liquid water not far under its surface today? Although the answer is still uncertain, it does appear to be possible. Given the incredible variety of places on Earth where life exists in the presence of liquid water, such as miles underwater at the bottom of the oceans, in hot springs, and in liquid water under the Antarctic ice sheet, it is natural to anticipate that life may have existed on Mars and perhaps still does exist inside it.

Of course, the speculation doesn't stop there. Strong evidence suggests that the interior of Jupiter's moon Europa, among others, is filled with liquid water. Does life exist there, too? Moving outward, the presently open question arises of whether there are worlds with liquid water orbiting other stars. Driven in part by such mysteries, the search for planets of any kind orbiting other stars has blossomed over the past several decades. The discovery of extrasolar planets actually began in 1992 and the number is rising through the hundreds (e.g., by the end of November 2009, 405 were known). Most of these are giants, like our planets Jupiter and Saturn, which are unsuitable as habitats for life.

Some of the scenarios in this book are based on the numbers and locations of extrasolar planets. One of the most intriguing related astronomical discoveries is the existence of Jupiterlike planets orbiting more closely to their stars than Mercury is to the Sun. We use this discovery in justifying one of our alternative Earths. Prior to these observations, astronomers did not believe it possible for giant planets to be even as close as Mars is to the Sun.

Planets are so small compared to the stars they orbit that until very recently all discoveries of extrasolar planets were made by detecting their gravitational effects on their stars (e.g., they make stars wobble as seen from Earth), rather than by observing the planets themselves. Planets have so little mass compared to the stars they orbit that only the most massive planets, much more massive than the Earth, have been detected. This is all changing. Technology has improved to the point where astronomers are beginning to discover Earth-sized planets orbiting other stars. While the holy grail of this realm of science is to find extraterrestrial life, Earthlike planets orbiting where water can be liquid on their surfaces are at least the grail holders.

The alternative scenarios that you are about to explore exist on the narrow boundary between science and science fiction. In this realm of writing, as in science itself, even the smallest change to the natural world leads to unexpected and often bizarre consequences. Furthermore, while exploring these alternative versions of Earth I came to really appreciate how making one change leads to myriad others. For example, changing the Sun's mass changes its temperature, which changes the distance from the Sun that a habitable world would have to be, which changes the length of the year, which changes the length of the seasons, which changes the challenges that life would face in such seasons, which changes the way species evolve, which changes . . . At the same time, that change in the Sun's mass also changes the intensities of the different kinds of radiation it emits, which changes the kind of protection life would need from, for example, ultraviolet radiation compared to what we need on Earth, which changes the concept of skin, which changes the way creatures move, which changes. . . . You get the idea.

I have made every effort to respect all the laws of nature in this process. As a result of the complexity intrinsic to this work, I cannot possibly explore every facet that is different from Earth on all the worlds in this book. I am sure that many of you will find other implications. I invite you to share them with other readers by posting them on the book's Web site, which can be reached through: http://us.macmillan.com/SMP.aspx.

Although I refer to scientific content introduced in other chapters, the scenarios in each chapter are independent of the ones in every other chapter. In other words, in each chapter I vary from scratch the environment in which the Earth was formed, rather than building on the alternative worlds created in the previous chapters. Some repetition between chapters occurs as I try to balance referring to the material that has been introduced earlier with the usefulness of presenting concepts in the context in which they are being used in various chapters. Also, all the units, such as day and year, refer to times as measured on Earth except when I explicitly specify where that time interval occurs, such as a Martian year or a Venusian day.

To help you distinguish among the worlds in the different chapters, I have given most of the bodies in most chapters different names. Each chapter is preceded by a short fictional vignette in which I envision how the changes made affect the lives of a few people living there. (Bear in mind that although I call them "people" and give them the same features we have to simplify the stories, the reality is that advanced alien life is extremely unlikely to look like us.) We begin by exploring the consequences of the Earth with two moons.

1 What If the Earth Had Two Moons?

DIMAAN, LLUNA, AND KUU

They came in the dead of night. One moment the bedroom was filled with the sounds of forest animals behind the village, the next with the deep shouts of men, rattling of heavily armored horses, and the ominous creaking of a carriage. A whip snapped; dogs began yelping; the leader of the group issued orders. They stopped right in front of his cottage. Lying in bed he heard his neighbors shuttering their windows and barring their doors. He knew that such actions would not protect them. The intruders would get in anywhere they wanted. As he finished this thought, his front door flew off its hinges, landing on the floor with a brittle crash.

He was up now, but as he moved to get out of bed, his mistress grabbed his arm. She was trembling, her eyes wide with the same terror he felt, but which he hid so she would not fear the worst. "Why?" she asked, hoarsely, softly.

"I don't know. Maybe," he smiled wanly, "I forgot to pay our taxes."

"They wouldn't be here for that."

"It was a joke," he said, lightly. She gave him the "not now" look, but he missed it as the stairs were filled with the clattering of hobnailed boots. Doors opened and slammed shut. The children began crying.

Then their bedroom door opened and an officer, followed by four soldiers, strode in.

"Galileo Galilei?" the officer demanded.

Galileo nodded. Without taking his eyes off Galileo, the man issued an order. "Take him."

"You will leave her and the children?" Galileo inquired, meekly.

"We only have orders to arrest you," the officer said, adding, "in the name of the Holy Inquisition."

If there is any justice in the world it occurred then, as a cluster of meteorites burst through the roof. A pair of them plunged into two of the soldiers, who fell dead.

Galileo laughed. The officer cursed and said, "Damned nuisance." Then he turned to the remaining soldiers and ordered them to take Galileo. Marina screamed as they dragged him out of bed. Galileo watched her, and their arms reached out to each other and then separated in slow motion.

The prison carriage, merely a cage on wheels drawn by a sickly horse, clattered and clanged as it carried the nightshirt-clad Galileo through the town and up to the castle. His body glowed red in the light of the moon Lluna shining and spurting molten rock overhead. He saw the eyes of hundreds of people watching him through slits in their shuttered windows. For perhaps the first time in his unruly life he wondered what they were thinking.

Over the next two weeks Marina, disguised as a scullery maid, brought Galileo his food, passing it through a small opening in the locked door, talking to him in whispers. She told him about the children and how the inquisitors had taken all his papers. He asked if "they" had asked her about anything in particular that he was working on. She shook her head.

For six weeks, Galileo sat in his prison room on a bed of straw and rough burlap, wearing the same nightshirt, which literally rotted on his body. During the first week, his arrogance kept him aloof, as he waited for inquisitors to question him. They never came. During the second week, his reserve turned to anger. The people taking turns monitoring him through a series of hidden mirrors saw him circling around the room, rubbing against the wall opposite his bed, then against the wall with the door, then against his bed, and finally against the outer wall, with the commode and window. As he circled, he rubbed the walls with his hand until it was raw. After the second week, Marina stopped coming. She could not stand the smells.

During the third week, as his anger dissipated, he found solace in continuing his studies. The day before his imprisonment, he had received a letter full of technical details of the observations made by Martinelli on the island of San Salvador in the New World. He had spent that day memorizing the results in it. Now he compared those observations, meticulously recorded by his friend, with the ones he had made at the same time and the ones that they both had made twenty years before. Sifting through all this data chiseled in stone in his eidetic memory, and scribbling calculations in the dust on the floor, he completed the work that had obsessed him for years. And he was right! The angles between the telescopes observing the same place on the moon Lluna had changed over the decades. Lluna was moving away from his world, Dimaan. Despite the squalor, the arrogance returned.

During the fourth week he started pounding on the door, demanding to be released. No one came to tell him to stop, and the door never opened. During the fifth week, confusion surrounding his arrest, imprisonment, and this interminable isolation broke through his defenses. The original question returned. Why had they arrested him? It must be something to do with his observations of the heavens, he reasoned, but what? His renowned ability to focus and concentrate evaporated.

During the sixth week, he started thinking about errors he had made: errors of commission and errors of omission. Maybe, just maybe, he should have married Marina. Perhaps he shouldn't have fired Sestilia. Vincenzo really deserved the raise he had requested all those years ago. And what about his girls....

The day after the tears appeared, an orderly opened the door, gagged, and vomited. Then he ordered Galileo out. Nearly naked, Galileo stumbled into the hallway. The gaggle of guards all backed away. With spears, they prodded him down the corridor and into a room through the middle of which ran a stream of water. He was tossed a piece of soap, a towel, and a robe, and ordered to bathe, which he did with as much zest as he could muster.

The courtroom was a study in contrasts. On the side where the three judges sat, the walls were covered with dark wood panels and an immense tapestry. Glasses and pitchers filled with crystal-clear water sat in front of each of them, along with baskets of fruits and nuts. On the other side, Galileo stood on a bare wooden platform in an alcove surrounded by gray, rough-hewn walls.

"How do you plead and do you agree to recant what you have said?" the central judge, tall, with a goatee beard, demanded.

"Are you out of your tiny mind?" Galileo demanded. "That is, if you have one at all. I have done nothing wrong. Nothing," he hissed at them.

The three judges and the guards against the side walls all gaped.

The judge on Galileo's right regained his composure first. Scribbling something on the sheet in front of him, he half turned his head toward the astronomer. "Do you really believe that? Do you think you would be here if we didn't have proof positive of your transgression?"

Galileo glared at him. The silence filling the room became so thick that several guards shook their heads to clear their minds.

"Do you deny..."

"I have done nothing wrong and made no mistakes," Galileo interrupted, through clenched teeth.

The third judge sat back, smiled briefly, and began speaking, putting his index finger to his lips as Galileo opened his mouth.

"We think that you misunderstand us." He motioned for the guards to leave. When they were gone, he rounded on Galileo. "We know that our planet, Dimaan, and the heavens have been here forever. They are immutable. Unchanging in Essential Essence."

"Earthquakes, volcanoes, sunspots," Galileo interjected, unsure where this was going.

"Mere challenges to humans," the judge said, smiling. "Our Creator does not want us to think we live in paradise here. You, however, claim that you can prove that there are irreversible changes in the universe. These, in turn, lead to the conclusion that our planet, and by extension, the universe, has not been here, fundamentally unchanged, forever."

"I...I don't follow," Galileo murmured, the bravado draining.

"Then I will explain...if you have a brain to understand," the judge sneered. "Dimaan has two moons, Lluna and Kuu. Lluna is nine-sixteenths the distance of Kuu, a relationship of perfect squares that the Church finds consistent with its teachings. You," he stabbed a finger at Galileo, "have secretly proposed that Lluna is moving away from Dimaan, toward Kuu. You have a co-conspirator, Luigi Martinelli, in San Salvador. He is there to provide you with data that will allegedly prove your hypothesis. And he has sent you that information."

The judge waved papers at Galileo, who squinted, unable to make out what was on them. The judge rose regally and, walking around the desk, handed them to Galileo. They were a copy of the letter he had received before his arrest.

"I don't unders—," he began, but suddenly he did. "This data, with my own, proves that Lluna is moving away. Hence it was closer yesterday and closer still each previous day. Once upon a time it must have been captured or it was part of Dimaan, flung off," he did a quick calculation in his head, "nearly a billion years ago."

"So if what you propose is true, Dimaan and Lluna were not the same in Essential Essence a billion years ago as they are today."

"Doesn't it give you a headache to think that everything has been as it is forever? What happened before forever?"

"Your caustic blasphemy will gain you nothing," the central judge said, his voice icy.

"If I can prove Lluna's recession, then science can rethink the evolution of our world. Otherwise, we are stuck with a universe that has lasted forever, a Lord who has also existed forever, and life that has only recently been put here. But why now? Why were we not created a trillion years ago? Or a billion years from now? It makes no sense to me."

Silence hung in the air like molten lead. "Our Lord works in mysterious ways. She has given us Her teachings and we are here to make sure that no one...I mean no one...goes astray. You, sir, are going astray and you are beginning to take others with you. This is absolutely unacceptable. You will...you must recant these heretical beliefs here and now and vow to never mention them again, except as errors."

"And if I refuse?"

"The past six weeks of confinement will only be the beginning. We will tortu…teach…you using every tool at our disposal. They are many tools, each more—instructive—than the previous one. Halfway through your instruction, you will plead for death—but it will be denied. Eventually we will rip your arms…"

"And what if…what if I am right and you are wrong? What if I can prove what I claim is true?"

The three judges looked at each other and the one on Galileo's right nodded slowly. "I think it would be a mistake to start with your 'education.' We will begin by educating your mistress and children."

E arth is unique in the solar system* for many reasons. Some distinctive properties are entirely obvious (complex surface life comes to mind), whereas others are more subtle, such as Earth having one Moon. In comparison, Mars has two moons, Jupiter has at least sixty-three, and the other planets have numbers between these two. (Venus and Mercury have none.) Furthermore, our Moon† has $\frac{1}{81}$ the mass of the Earth, whereas all the other moons in the solar system have masses less than $\frac{3}{10,000}$ the masses of their planets.

The Moon's existence, combined with its large mass compared to the Earth and the fact that it is the only natural body orbiting our planet, has led to many changes from what the Earth would have been like without it. Because of the Moon's existence life formed relatively rapidly; the day is twenty-four hours long rather than being roughly eight hours long; tides are three times higher than they would be otherwise; many species of animals that are active at night could not exist without moonlight to aid their hunting, navigating, and mating activities; our planet's rotation axis does not randomly change direction as it would otherwise do; and we have an essentially constant cycle of seasons, which we wouldn't otherwise have, among many other things. Each of these results of our having a Moon, in turn, has affected myriad other aspects of the Earth and life on it. Any modification to the Earth or its astronomical environment leads to fascinating changes in

---

* The solar system is the Sun and everything that orbits it, namely the planets, moons, and a variety of debris.
† Following astronomical convention, moon is only capitalized when referring to our Moon.

our world, as well as providing new perspectives and insights into our planet as it is now. In this first of ten alternate worlds we explore what Earth would be like if we had two massive moons today instead of just one.

The Earthlike planet in this chapter, called Dimaan, begins its life identical to the early Earth in size, composition, and distance from the Sun. Based on geological and fossil evidence, the Earth was initially spinning much faster than it is today. Although that rate is not yet known, I give Dimaan a plausible eight-hour day when it first formed. Neither Earth nor Dimaan had a moon at first. Ours came into existence within about 200 million years of the Earth's forming.

Moons can form in four ways: from impacts, in which the planet is struck and thereby ejects debris that becomes one or more moons; simultaneously with a planet, in which the moons and planet condense together (Appendix); by fission, wherein the moons are literally thrown off a rapidly rotating planet (Appendix); and by capture of the moons after the planet has formed.

Most astronomers believe that our Moon formed as the result of a collision between Earth and a Mars-sized body. The intruder hit Earth at an angle that ejected debris into orbit in the same general direction in which our planet was spinning. This rubble formed a short-lived ring that was much smaller but, interestingly, much more massive than all of Saturn's rings combined. As this material orbited, it began colliding with itself and bunching together under the influence of its own gravitational attraction until it coalesced into the Moon. This is how I posit Dimaan's first moon, Kuu, formed.

Although it is entirely possible for an impact of a small planet onto a larger one to splash enough debris into orbit to form two moons similar to ours, such moons would drift together and collide billions of years before advanced life evolved on Dimaan (Appendix). Because I want that second moon around for people to enjoy, I posit that Dimaan captures its second moon long after the first one formed.

CAPTURE OF THE SECOND MOON

The process of forming a star and its host of orbiting objects is a very, very messy affair involving countless collisions. The star system begins as a slowly swirling eddy of gas and dust in a giant interstellar cloud that begins to contract under the influence of its own gravitational attraction. In the case of our solar system or that of Dimaan, the central region of this eddy condenses to become the Sun. The material in its outer reaches becomes a disk of gas and dust, parts of which condense to form the planets, moons, and smaller orbiting pieces of debris such as asteroids and comets.

Most moons are potato-shaped bodies typically a few miles across that were originally not bound to planets. As these small bodies drifted past them, the planets captured them with relative ease. In our solar system, the tiny moons include Phobos and Deimos orbiting Mars, and at least 150 moons orbiting the giant planets Jupiter, Saturn, Uranus, and Neptune. For each piece of space debris that was captured, millions of similar objects struck planets or sped past too rapidly to go into orbit.

Even though our solar system and the one destined to become Dimaan's home began identically, it is entirely plausible that Dimaan acquired a second massive moon. This happened because the orbits of the debris from which all the bodies in the solar system formed were chaotic, a concept that transcends the intuitive definition of the word.

The mathematical and physical concept of chaos reveals that exceedingly tiny changes (such as a butterfly flapping its wings in Africa) can have monumental consequences (such as a hurricane in Louisiana) that would not have occurred without the tiny initiating event. In the formation of the solar system, tiny gravitational tugs from small pieces of debris led to huge unpredictable changes in the orbits of all the objects that formed in it, compared to what the orbits would have been had the small pieces not been there. For example, imagine two mountain-sized chunks of rock that gently collided in the young solar system, thereby creating a slightly larger object. This bigger body then collected other matter, eventually growing into the Earth. Now suppose that the initial collision was ever-so-slightly faster so that the impact pulverized the two bodies, rather than forming a larger, more massive one. In that case, the Earth would not have formed as it did. Another collision of different debris, perhaps at a different distance from the Sun, would have led to the formation of an Earthlike planet, but with a different orbit and different physical properties than the Earth has today.

Chaos also justifies the assertion that if another system very similar to the solar system had formed with Dimaan in the same orbit as Earth, there could easily be a few "extra" Moon-sized bodies in it that never existed in our solar system. One of these is destined to be captured and become Lluna, Dimaan's closer moon in this chapter.

Let's now set the stage for Lluna's capture by considering how Kuu evolved into its present orbit. Once we have gotten it out there, we can capture Lluna. Because Kuu and our Moon formed identically, I simplify the discussion by referring here to how our Moon got out to its present orbit.

If the impact that splashed the debris that became our Moon into orbit had been powerful enough to put the Moon directly into its present orbit, the Earth would have been destroyed in the process. Keeping the Earth intact required that the debris splash into orbit much closer to the planet. We don't yet know how near to the Earth the Moon was originally, but it could plausibly have been ten times closer.

Tides are the key to understanding how the Moon got out to where it is today, along with why the day is twenty-four hours long. Indeed, the tides play roles in so many of the scenarios in this book, I believe it is essential that we briefly explore them here.

Tides

For the first twenty years that I taught astronomy, I explained the tides incorrectly. The explanation in the existing astronomy texts back in those days, even the advanced ones, from which I had learned about the tides, was wrong. It went like this. The force of gravity decreases with distance (which is true). The part of the Earth closest to the Moon feels the greatest gravitational attraction to it, and the central region of the Earth feels less attraction. Because the ocean water closest to the Moon is pulled hardest, that water creates a high tide. The oceans on the far side of the Earth feel the least gravitational attraction, so they are "left behind," meaning that they create a high tide on the side of the Earth farthest away from the Moon.

My culpability in propagating this misconception runs deeper. During that period I began writing an astronomy text that included this explanation. Our understanding of the cosmos changes rapidly, so astronomy texts are revised every few years. Each time we revise one, it goes out to many astronomers for review. Eventually one reviewer (out of over a hundred) complained about the above explanation of tides. Needless to say, I thought he was wrong, but in the spirit of "due diligence" I decided to check. I went to the font of much wisdom about the oceans, the National Oceanographic and Atmospheric Administration. They have a Web site entitled *Our Restless Tides* that convinced me that there is more to tides than I and my colleagues had been writing and teaching.

After reading this, I invite you to let me know whether you had been given the correct scoop on tides by visiting this book's Web site.

Although I describe tides for the Earth–Moon system, this presentation applies to any objects on which tides occur. To simplify matters, let's begin by ignoring the Earth's spin (technically, rotation) on its axis and the tidal effect of the Sun. We put them in later.

There are two parts to the tidal picture. First is gravity, the only universal force of attraction. Although the effect of gravity from any object extends infinitely far, the strength of the gravitational attraction decreases with distance: the farther you are from something, say the Moon, the less gravitational force you feel from it. Consider, then, the Moon's gravitational attraction on the Earth's oceans. The oceans closest to it at any moment feel the strongest gravitational attraction from it. Those halfway around to the other side of the Earth feel less force from it, whereas the oceans on the far side of the Earth feel the least gravitational attraction from it. This change in attraction of the oceans by the Moon with distance is not enough to explain the tides.

The second piece of the tidal puzzle relates to the orbit of the Moon. Contrary to popular belief, the Moon does not orbit around the Earth. Rather, the Earth and Moon together orbit around a common point, their center of mass, like two dancers who are holding each other and waltzing around. When gliding straight across the dance floor, the dancers whirl around each other and their center of mass moves in a straight line.

The center of mass of the Earth–Moon system is called the barycenter. It is located 1,064 miles under the Earth's surface on a straight line between the centers of the two bodies. The barycenter follows a smooth elliptical orbit around the sun, whereas the Earth and Moon waltz around it (Figure 1.1a).

The motion of the Earth around the barycenter creates a force everywhere on the Earth that is directed straight away from the Moon. This is similar to the outward force you feel on a merry-go-round. The tides result from a combination of the gravitational force from the Moon pulling the oceans toward

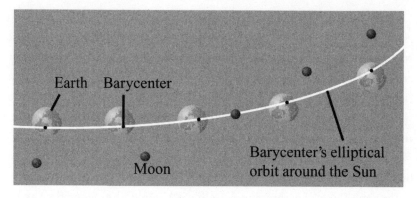

Figure 1.1a: Earth and the Moon orbit around a common center of mass, the barycenter. This point moves in a smooth elliptical orbit around the Sun. CREDIT: W.H. FREEMAN & CO.

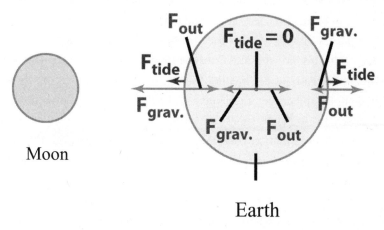

Figure 1.1b: The tidal force, F_{tide}, is created by subtracting the gravitational force, F_{grav}, from the outward force, F_{out}, which is created by Earth's motion around the barycenter. These forces are shown at three places on Earth. CREDIT: W.H. FREEMAN & CO.

it and the force away from the Moon created by the motion of the Earth around the barycenter. Keep in mind that the force of gravitation decreases with distance and the outward force is constant over the Earth. When you subtract these two effects everywhere (Figure 1.1b), you discover that the closest point on the Earth to the Moon feels a net (i.e., gravitational minus outward) force toward the Moon, whereas halfway around the Earth to the other side the net force is zero, and on the opposite side of the Earth from the Moon, the net force is directed away from the Moon.

Aha! you say. The net force toward the Moon lifts the waters closest to it, and the net force away, on the opposite side of the Earth, pushes those waters away from the Moon. Not quite. If those net forces were the only cause of tides, then tides could only be a matter of inches high, because that is the maximum height that the Moon's gravitational force can lift ocean water. What really causes the tides is that the water nearly halfway round to the other side from the Moon is pulled by it, sliding along the Earth's surface, piling up close to the Moon, and causing the oceans closest to it to bulge outward (a high tide). At the same time, the tides just beyond halfway to the opposite side are thrust away from the Moon, sliding along the Earth's surface, and piling up opposite the Moon in another, simultaneous, high tide as shown in Figure 1.1c.

The above discussion implies that the high tide closest to the Moon should be in a straight line between the centers of the Moon and Earth. It isn't. We need to put the Earth's rotation back in the mix. As the Earth turns, it pulls the high tides around with it. As the one closest to the Moon is pulled away from being directly under it, the Moon pulls it backward, trying to keep that tide lined up between the centers of the two worlds. Because the Earth spins nearly twenty-eight times faster than the Moon appears to orbit the Earth, our planet's motion combined with the friction between the ocean and the surface of the Earth keep that high tide about ten degrees ahead of the Moon (Figure 1.1d).

Finally, as concerns the tides, it is worth noting that although the Sun is about twenty-seven million times more massive than our Moon, the Sun is 390 times farther from the Earth. The relatively short distance between the Earth and Moon, compared to the distance between the Earth and Sun, causes the tidal effect of the Moon to be greater than the tidal effect of the Sun. Today, the Moon creates two-thirds of our tides and the Sun creates most of the rest.*

If the high tide closest to the Moon were directly between the centers of the Moon and Earth, the gravitational attraction of that water on the Moon would pull it in the same direction as the rest of our planet does, keeping

---

* Jupiter and the other planets create only a very tiny fraction of the tides here because the gravitational attractions of all these bodies for the Earth are well below the attractions of the Moon and Sun.

Figure 1.1c: *Displacement of the oceans due to the tidal force from the Moon.* CREDIT: W.H. FREEMAN & CO.

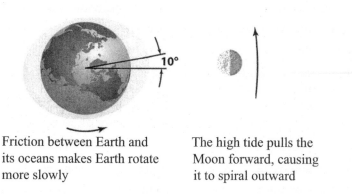

Friction between Earth and its oceans makes Earth rotate more slowly

The high tide pulls the Moon forward, causing it to spiral outward

Figure 1.1d: *The high tides are displaced by about ten degrees due to the Earth's rapid rotation.* CREDIT: W.H. FREEMAN & CO.

the Moon in its present orbit. Because the closer high tide is not on that line, the water in it creates a gravitational attraction on the Moon that pulls it forward (Figure 1.1d), giving the Moon extra energy, which causes it to spiral outward. With this as background, we can now return to the evolution of the moon Kuu around Dimaan.

Forming ten times closer than the Moon is to the Earth today, Kuu created tides on Dimaan that were 1,000 times higher than the tides we now experience from the Moon. These towering tides, pulling Kuu forward in its orbit, provided the tug necessary to cause that moon to spiral outward. After about 4.2 billion years, Kuu was out nearly as far as our Moon is today.

Life began on Dimaan when it had just the one moon, Kuu. Aquatic plant life in Dimaan's oceans evolved first. It was just when plants were about to transition onto the land, about 4.2 billion years after Dimaan and Kuu formed, that Lluna began drifting toward it.

Lluna, a spherical body similar in size and mass to Kuu, formed far from Dimaan. Not orbiting a planet, we call such bodies asteroids or "small solar system bodies." In its early years, Lluna acquired a companion body $\frac{1}{10}$ the mass, like our dwarf planet Pluto and its main moon, Charon (pronounced like the name Karen). Indeed, many small objects in our solar system have orbiting companions. Lluna's companion will play a crucial role in the moon's capture by Dimaan.

Capture of massive moons isn't easy. Gravity pulls objects together, but a near miss doesn't imply capture. If the incoming body or bodies are moving too fast, they will slide by, changing direction as they go, and then recede back into interplanetary space. To be captured, an object from afar must slow down or, put technically, it must lose the energy it has as a result of its motion, called kinetic energy.

Kinetic and Potential Energy

Two kinds of energy must be considered in understanding how one body can capture another: kinetic energy and potential energy. Imagine jumping upward, rising perhaps a foot. The energy you have when you are moving is called kinetic energy. Once you leave the ground, of course, you begin slowing down, which means that your kinetic energy is decreasing. The Earth's gravitational attraction (or force) is what slows you. In situations like this, however, energy is conserved, meaning that your kinetic energy has to be converted into another kind of energy in your body. Your kinetic energy is transformed into potential energy, which gives you the "potential" of falling back down. Assuming that nothing prevents that from happening, your potential energy is converted back to kinetic energy as you descend.

This kinetic energy/potential energy interplay occurs for all objects moving through the solar system. Consider the object destined to become

Lluna. Far from Dimaan, Lluna and its companion are traveling along, feeling the planet's gravitational pull. Therefore, Lluna has both kinetic and potential energies. As it nears Dimaan, the force between them increases, the kinetic energy of Lluna increases (it speeds up), and its potential energy decreases. After the closest approach, Lluna and its companion recede and slow down the same amount they had originally sped up. In other words, they leave the vicinity and don't necessarily ever return. If that were all there were to it, Dimaan could never capture Lluna because Lluna must lose overall energy (i.e. kinetic plus potential) to be captured. If we can figure a way for Lluna to lose energy, it will be unable to move as far outward as it was originally. In other words, it would be captured. This is the same as if you jumped upward a second time, but with less vigor (think, energy). In that case, you would not rise as high.

The Capture

Four effects contribute to the capture of Lluna: most important is the fact that Lluna's companion feels a slightly different gravitational attraction from Dimaan and Kuu than does Lluna itself. This occurs because as Lluna and the companion approach Dimaan, these two intruders are at slightly different distances from the planet and its original moon. Therefore they feel different amounts of gravitational force from them. This difference can be enough to pull the companion free of Lluna and fling it away, taking with it a substantial amount of energy, which has the effect of slowing Lluna down, making it possible for the final three effects to complete its capture.

Upon approaching the Dimaan–Kuu system, Lluna's gravitational attraction pulls on the moon Kuu, causing its orbit to become more elongated (more elliptical). Moving Kuu causes Lluna to lose energy. At the same time, Lluna creates tides on the planet Dimaan that pull back on Lluna, slowing it down further. Finally, the gravitational pull of Dimaan on Lluna coupled with the planet's orbit around the Sun cause Lluna to lose even more energy. In this final process, energy is taken from Lluna and given to Dimaan. The combination of all these effects can remove enough energy from Lluna for it to become locked in orbit around Dimaan.

I set Lluna's initial orbit around Dimaan to be half Kuu's distance from the planet, with both moons orbiting in the same direction and in about the same plane that our Moon orbits Earth. As we show shortly, this leads to eclipses related to both Kuu and Lluna. Virtually all objects in the solar system have elliptical orbits (egg-shaped), however, most of these are very close to circular. Lluna and Kuu will initially have more elliptical orbits than any other moons because the capture of Lluna was so messy.

It will take roughly two weeks from the time that Lluna is first close enough to generate noticeable tides on Dimaan until this moon is securely in orbit. During that transient period, all hell breaks loose on the planet. Lluna's gravitational pull creates tides on Dimaan eight times higher than those from Kuu. While Lluna is settling into orbit, it will also create monster tidal waves on Dimaan that will make any tsunamis that we have on Earth seem like tiny ripples in comparison. The water will slosh like the waves created in a large pan filled with water as you carry it from the sink to the stove.

These tidal waves and the tidal bulges generated by Lluna will create Dimaanquakes and increased volcanic activity that will persist for years. The dust released by the volcanic emissions will darken the skies and cool the atmosphere dramatically. The volcanoes active during this time will also release vast volumes of water vapor, carbon dioxide, sulfur dioxide, carbon monoxide, stinky hydrogen sulfide, and hydrochloric acid, among other gases. All of this activity will cause a mass extinction in the ocean life of Dimaan.

LIFE WITH LLUNA

Lluna's capture and the damage to Dimaan and life on it in the process don't mean that the planet will thereafter be lifeless. Life on Earth has experienced over half a dozen similarly catastrophic mass extinctions, episodes caused by geological and astronomical events during which large fractions of all life on our planet were eradicated. Perhaps the most dramatic of these events, the Permian–Triassic extinction, occurred 251 million years ago. It wiped out over ninety-five percent of all species of life. Nevertheless, the remaining life-forms grew, diversified, and became the progenitors of the life on Earth today. What Lluna's presence does mean is that the sequence of evolutionary events on Dimaan would be profoundly different from what occurred here on Earth or that would occur on Dimaan had Lluna not appeared on the scene. Let's explore some of the differences that would result.

Llunalight

At half the distance, Lluna will have twice the diameter as does Kuu in Dimaan's sky (or does our Moon in our sky). Twice the diameter means that the area Lluna covers in Dimaan's sky will be four times greater than that of Kuu. Because moonlight is sunlight scattered from the surface of a moon, Lluna will be four times as bright on Dimaan as is Kuu.* Combining the light from both moons, nighttime on Dimaan when both moons are full will be five times brighter than the nighttime surface of the Earth under a full Moon. It would be quite easy to read a book under those conditions.

Lluna and Kuu orbit Dimaan at different speeds, therefore it is more likely that at least one of the moons is up at night than it is for us with our single Moon. When a moon is high in the sky at night it is at least half full (technically the moon is in either a gibbous or full phase). Therefore, Dimaan will have more nights brightly lit with moonlight than does the Earth.

In what follows, let's assume that the sensory equipment available to life on Dimaan is the same as on Earth. That means people there will evolve seven senses: touch, taste, smell, sound, sight, heat, and gravity. The last two of these are often left off lists of senses taught to children, but we have them nevertheless. Sensitivity to heat is straightforward: put your hand near a fire and you know that it is hotter than its surroundings. Sensitivity to gravity is our ability to know our posture and to sense when we are falling.

Because it will be easier for predators to see their prey at night on Dimaan, camouflage will be more highly refined than it is on Earth. This, in turn, will require more acute hunting skills using sight, sound, smell, and heat detection for animals that are active at night. The cycle of protection and detection driven by the brighter nights on Dimaan could well lead to creatures that are more aware of their surroundings than early land animals

---

* This assumes that their surfaces are made of the same materials. This is not always true and different moons of the same size in our own solar system have very different intrinsic brightnesses.

were on Earth. This, in turn, is likely to increase various aspects of intelligence in these creatures compared to what was necessary for survival here. Perhaps the first sentient creatures on Dimaan will evolve from nocturnal hunters rather than from arboreal creatures, as occurred on Earth.

Eclipses

Eclipses happen when the shadow of one world falls on another. Solar eclipses on Earth occur when the shadow of the Moon crosses the Earth when the Moon is in the new phase (between the Earth and Sun). Conversely, lunar eclipses occur when the full Moon moves into Earth's shadow. However, you have probably noticed that solar eclipses do not occur at every new Moon, nor do lunar eclipses occur at every full Moon.

Eclipses don't happen every month because the Moon's orbit is tilted five degrees compared to the plane of the ecliptic (Figure 1.2). The ecliptic is the plane defined by Earth's orbit around the Sun. When the Moon is new, it is usually slightly above or below the ecliptic, meaning that the Moon's shadow is slightly above or below the Earth. When our Moon is crossing the ecliptic in the new phase, however, its shadow speeds across a swath of the Earth, prevents sunlight from striking this region, and thereby creates a

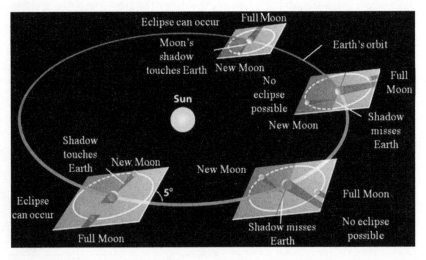

Figure 1.2: Eclipse conditions for Earth and our Moon. The large oval, the path of the Earth around the Sun, defines a plane called the ecliptic.

solar eclipse. The same arguments about the Earth's shadow apply to lunar eclipses when the Moon is full.

Lluna does not cause solar eclipses every time it passes between the Sun and Dimaan because I put Lluna in an orbit that is also tilted 5° from the ecliptic. Lluna is only 1° across as seen from Dimaan. Therefore, when Lluna passes between Dimaan and the Sun, its edge closest to the Sun will appear as much as 3¾° from it. At such times, Lluna is too far in angle from the Sun to cover it as seen from the planet.

When solar eclipses on Dimaan caused by Lluna do occur, they will be different from those caused by Kuu (or, equivalently, by our Moon on Earth) in two ways. Total solar eclipses occur when a moon's shadow touches its planet. If you are inside that shadow, the moon would appear dark (new phase), with the Sun completely blocked out behind it. On Earth that complete shadow region behind the Moon, called its umbra, is never more than about 160 miles wide on the Earth's surface. If you are just outside the umbra, in the penumbra, you will see a crescent Sun on the edge of the Moon throughout the eclipse. The umbra sweeps across the Earth's surface so rapidly that an eclipse never lasts for more than about seven and a half minutes here.

The first difference for Lluna's eclipses is that its larger apparent size in Dimaan's sky will cause its umbra to cover twice the area on the planet's surface as does our Moon's umbra. The second difference is less obvious: the total solar eclipse caused by Lluna lasts a shorter time than such eclipses do on Earth today. This difference in duration occurs because Lluna orbits closer to Dimaan. Kepler's third law of orbital motion, worked out by Johannes Kepler (1571–1630), reveals that the closer a moon is to its planet, the faster it orbits compared to a moon farther away. Hence Lluna moves faster across the sky than does Kuu. Its higher speed more than makes up for the larger size that Lluna has in Dimaan's sky as the shadow passes across Dimaan.

Phases of the Moons

The two moons will have cycles of phases that last different lengths of time as seen from Dimaan. The phases of a body are caused by seeing different amounts of its sunlit side from hour to hour or day to day. From Earth, for example, we see varying amounts of the lit side of the Moon as it orbits around us. Starting when the Moon is between the Earth and Sun, this cycle of phases changes as follows: new (when we see the least of the lit side), waxing crescent, first quarter (when we see half the lit side), waxing gibbous,* full (when it is on the opposite side of the Earth from the Sun and we see all the lit side), waning gibbous, third quarter, waning crescent, and back to new. This cycle takes about 29½ days to complete.

Lluna and Kuu will both go through the same order of phases. When Lluna is first captured, Kuu's cycle will have essentially the same length as our Moon's phases, namely about 29½ days to go from, for example, new moon to new moon. Lluna, orbiting more rapidly, will have a more rapid cycle of phases that takes about 10 days to complete. That cycle is so rapid that you could go out and see Lluna in one phase and return later that day or night and see it in a distinctly different phase!

We see the phases of our Moon vary uniformly because the Moon's orbit around us is nearly circular. Because they have significantly more elliptical orbits, the phases of Kuu and Lluna will change much less smoothly. When they are each closest to Dimaan, their phases will be changing most rapidly, and conversely the phases will flow most slowly when they are farthest away.

Tides and Shorelines

With a few notable exceptions, such as Swansea, Wales, and Mont Saint Michel, France, where the ocean bottom slopes downward especially slowly, beaches on Earth tend to vary in width by a few hundred feet or less between high and low tide. Although tidal erosion is significant, it usually takes place over periods of decades or longer, giving people who live or work on the shore time to develop defenses against the changing landscape. The presence of Lluna will make both the range of tides and the speed at which they erode the shoreline of Dimaan much greater.

As noted earlier, Lluna creates tides 8 times higher than Kuu. Combin-

---

* Gibbous phases are those where we see more than half, but less than all, of the Moon. These phases are distinguished by humps of unequal shape on opposite sides. On the other hand, during crescent phases, when we see less than half of it, one side curves outward and the other side curves inward.

ing their tidal effects with that of the Sun leads to tides on Dimaan that are as much as 6⅓ times greater than the range of tides on Earth today. This occurs when Dimaan, Lluna, Kuu, and the Sun are in a straight line. Therefore, the typical intertidal region (the area that is exposed at low tide and hidden at high tide) on Dimaan will be much more extensive than it is on Earth. Because more tidal water is flowing on Dimaan each day, the amount of erosion of the shoreline there will be much greater than it is here. The shore would therefore wear away and expose coastal buildings to damage more rapidly than occurs on Earth.

Tidal erosion would also be more profound in the rivers on Dimaan than on Earth. Where the rivers run into the oceans on Dimaan, the tides would carve deeper channels than they do here. These would lead to tidal waves called tidal bores that even run up some rivers here on Earth, such as those off the Bay of Fundy between Maine and Nova Scotia, and the Bristol Channel between England and Wales.

Cities built near the mouths of rivers, such as New York, San Francisco, New Orleans, or even London (as far upriver on the Thames as it is), would experience unacceptable erosion problems due to the tides and tidal bores generated on Dimaan. Shorelines would erode so quickly that without heroic effort, such as thick concrete walls lining the rivers, cities could not be built on most ocean shores or on especially active rivers.

It is worth noting that building cities on waterways has the advantages of providing easy sewage removal and of transporting people and goods to and from many locations via ships. It is therefore likely that the challenges caused by high tide cycles would lead to differences in the way coastal civilizations developed on Dimaan from the way they have on Earth.

Calendars/"Months"

The month on Earth has its origins in our Moon's cycle of phases. That cycle takes roughly 29½ days. The fact that this is not a whole number of days, and especially not a whole number divisible by seven (for the days in the week) has made the month less useful and more symbolic than the day, week, or year.

Dimaan's calendar could conceivably be even more convoluted, because

Lluna and Kuu have different cycles of phases and the two cycles will not be whole number multiples of each other. Dimaan's people could plausibly have calendars with days, weeks, parmos,* and months.

Lluna's Rotation (and Its Change Due to Tides)

As with all similar-sized bodies in our solar system, Lluna will be rotating when it is captured by Dimaan. The rate at which it rotates will be unrelated to any property of Dimaan: Lluna's rotation will have been established long before, when it and its companion were orbiting around each other. In our solar system, the dwarf planet Pluto rotates once every six days which, not coincidentally, is the same length of time it takes Charon to orbit around it.

Charon, our Moon, and Dimaan's moon Kuu all have an interesting orbital property, namely that they rotate at exactly the same rate as they orbit their planets. This is called synchronous rotation. Objects in synchronous rotation always have the same side facing the body they orbit. That is why we always see the same side of the Moon.

Synchronous rotation of spherical moons is not an accident. When moons form or are captured, they are unlikely to rotate at the same rate they orbit. However, the planet creates tides of the land on their spinning moons called land tides. These are analogous to the tides that moons create in oceans. As land tides cause the ground to rise and fall, friction inside the moon between adjacent pieces of moving rock creates heat (as you would feel if you rubbed your hands together vigorously). This heat, in turn, melts that rock, creating molten rock—magma—that leaks out through volcanoes and cracks in the moon's surface. This liquid rock flows on the surface, enhancing the tidal effect.

This idea of land tides is not unique to Lluna, nor does the rock that moves have to be molten to be tidal. Surprisingly for those of us who are not geologists, Earth's continents have tidal motion today (commonly called Earth tides, but that name will confuse matters later on). They were proposed in the 1880s by Sir George Howard Darwin and measured as early as 1909. The highest ones here, over eight inches high, occur in the equatorial regions of our planet

---

* For "partial months," the length of Lluna's cycle of phases.

as the land is pulled upward by the gravitational attraction of the Moon and Sun. They can be measured using technology, but you can't feel or see them.

The regions of high land tides act as handles that the gravitational attraction of the planet can pull upon to change a moon's rotation rate. At the same time, friction between liquid rock flowing on the moon's surface—lava—and the solid rock below it changes the rotation rate in the same way. Eventually these two effects combine to change the moon's rotation rate from whatever it was to a rate equal to the time it takes the moon to orbit the planet. In other words, the moon's rotation becomes synchronous. At that time, the high tides on opposite sides of that moon become fixed (i.e., they will no longer travel along the surface) and so the rubbing that created heat before it was locked into synchronous rotation will cease. The surface then cools and solidifies.

Most of the moons in our solar system are in synchronous rotation around their planets. Lluna will be brought into synchronous rotation. However, another source of heating will keep Lluna's interior molten.

Volcanoes on Lluna

By far the most spectacular thing about Lluna's presence at the time people exist on Dimaan will be the moon's active volcanoes. Their existence is analogous to the volcanoes that occur on Jupiter's moon Io today. Recall that because of the way it was captured, Lluna's orbit around Dimaan is not especially circular. Eventually, its orbit becomes more so, but Kuu prevents it from ever being perfectly circular: when Lluna is between the planet and the outer moon (Figure 1.3a), Dimaan pulls it in one direction, and Kuu pulls it in the opposite direction. As a result, Lluna is pulled into an orbit slightly farther away from Dimaan than when Lluna is on the opposite side of the planet (Figure 1.3b). In the latter position, both the planet and the other moon are pulling it toward Dimaan and so Lluna is then closer to the planet than it would be if Kuu were not there.

The result of the noncircular orbit is that when Lluna is closer to Dimaan, the land tides on the moon are higher than when Lluna is farther away.* Seen

---

* Keep in mind that the moon is in synchronous rotation, so the high tides on the land will be facing toward and directly away from the planet as the moon goes around.

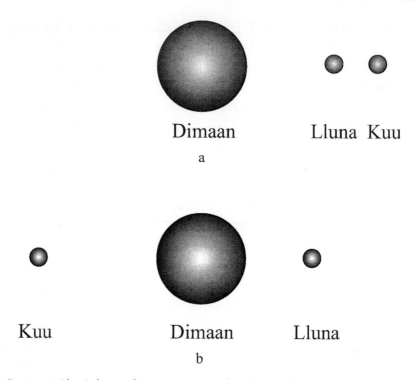

Dimaan

Lluna Kuu

a

Kuu

Dimaan

Lluna

b

Figure 1.3: Lluna's distance from Dimaan varies depending on the location of Kuu.

from afar, it would look as if Lluna were breathing as it orbits Dimaan. This change in height of the land creates the same friction that kept it molten before it was in synchronous rotation. As a result, the inside of Lluna will be molten throughout its existence in orbit around Dimaan and that magma will continually be leaking out through volcanoes and cracks in its surface.

Lluna is going to present a spectacular sight from Dimaan. Peppered with huge volcanoes, Lluna will be a world alive with red-hot lava being ejected in several places at once. Some of these events will be seen along the edge of the moon, like mammoth mushroom-shaped fountains leaping miles into the air and then crashing silently back down. These are analogous to stratovolcanoes on Earth. Other features on Lluna will include volcanoes that ooze lava, and rivers of lava that flow and, upon cooling, freeze into place. Although that would be very romantic today, I can imagine that prescientific civilizations on Dimaan would create a wide variety of mythical scenarios based on it. Hell, yes.

Unlike our Moon, there will be few, if any, impact craters on Lluna by the time people view it from Dimaan. Lluna's surface will be re-covered by fresh lava so often that craters will be quickly erased.

Some of the lava ejected from Lluna will be racing upward so rapidly that it will escape into space and never return. Some of that debris will drift away from the Dimaan system, some of it will form a tenuous ring around the planet, and some of it will hurl toward the planet.

Pieces of debris, pebble-sized and bigger, that are ejected Dimaanward from Lluna will continually be entering the planet's atmosphere. The smaller pieces will vaporize, creating meteors, just as occurs in our skies. Whereas we experience frequent meteors only a few times a year, during meteor showers, every clear night on Dimaan will be filled with meteors (aka shooting stars) streaking down every few seconds. That would be very romantic from our perspective. Would Dimaanians become so inured to these events that they would lose interest?

Larger pieces of debris falling through an atmosphere do not completely turn into dust. Their solid remains strike whatever is in their paths. Sufficiently large ones create craters upon impact, but the pieces of debris from Lluna are likely to be too small to do this. Nevertheless, life on Dimaan is going to have to contend with more frequent meteorite impacts, like bullets from the sky, than we experience on Earth.

The Collision Between Lluna and Kuu

Lluna and Kuu are destined to collide. Our Moon spirals away from the Earth. Kuu spirals away from Dimaan. After Lluna forms, the tides it creates on Dimaan will act back on it, forcing it, too, to spiral outward. Calculations reveal that after Lluna is captured, it recedes from Dimaan faster than does Kuu. Just as the recession of our Moon slows the Earth's rotation and Kuu's recession slows Dimaan's rotation, the recession of Lluna will make Dimaan's day even longer. As Lluna closes in on Kuu, the day on Dimaan will exceed twenty-eight hours.

The collision will take place tens of thousands of miles farther from Dimaan than our Moon is from the Earth. As seen from Dimaan, Lluna will approach Kuu from behind. In their final days, the gravitational forces that each moon exerts on the other will cause the two to become more and more egg-shaped as land tides miles high form on them. This will cause Lluna to crack open, allowing its molten interior to pour out and cover its surface in glowing lava. Kuu, heated by friction as it distorts, will also have an outpouring of molten rock. Then the worlds will kiss.

Inexorably Kuu and Lluna will come together. The collision, however, will not be even remotely as horrendous as the impact on Dimaan that created Kuu or the one on Earth that created our Moon. The intruders in these latter impacts were moving much more rapidly, compared to the planets they struck, than Kuu and Lluna will be moving when they collide. People on the side of Dimaan facing their moons can prepare a jug of their favorite libation, set up a comfy chair, and watch the first phases of the event over a period of hours, as they might watch a science fiction movie at home. Only this event will be real and its consequences deadly.

After the kiss, rings of crushed rock will fly off the moons from the regions where they come into contact. At the same time, the sides of the moons opposite the impact site will erupt with lava shooting out as their liquid interiors collide and bounce away from each other. By the time a quarter of their mass has come into contact, both moons will begin to break apart. Seen from Dimaan, they will appear to explode in slow motion. For many hours the sky will be filled with bright red light from the impact region and wherever else molten rock is emerging.

Debris from the impact will fly in all directions, most notably perpendicular to the direction that the two bodies were moving when they struck. Put in blunt terms, a lot of stuff is going to fly toward Dimaan. The devil of this impact is in the details (such as the relative speed between the moons, their internal temperatures, and their chemical compositions), but some of their debris, including chunks big enough to create craters miles across, will drift toward the planet for several days and then fall into Dimaan's atmosphere.

Fortunately, this event is going to occur long after advanced civilizations have been established on Dimaan. They are likely by then to have the technology and techniques (Hollywood, take note) to prevent impacts of collision debris that would otherwise lead to a mass extinction on their world. Whether they would be able to save the civilizations they had established on the two moons before the collision is another question altogether.

FINALE

The two moons will eventually become one. Following the impact, debris that wasn't blown completely out of orbit would form a significant ring around Dimaan. Within a few years, the densest part of the ring would clump together due to its own gravitational attraction and due to relatively slow collisions between its pieces. A single body would form, growing as it collected more and more of what used to be Lluna and Kuu. Eventually this new moon would absorb the remaining ring debris and a new era in the life of Dimaan would begin.

2

What If the Earth Were a Moon?

Mynoa, at sea in the middle of the Atlantic Ocean,
early October 1492.

Eight bells on the first watch. Christopher Columbus stood on the poop deck of the *Santa Maria,* holding on to the shrouds as his flagship pitched sickeningly in the choppy seas.

"Nothing, Captain," the lookout in the crow's nest shouted down to him.

"Keep a weather eye out there, Mendez," Columbus ordered. He looked back over the stern at Tyran, the planet around which his world orbited. Back home there was something reassuring about going out any time of day or night and seeing the magnificent turquoise orb fixed in the same place in the sky. Over the past month and a half as they sailed westward Tyran had settled slowly in the east.

"Mr. Gutierrez, we must deal with the mutineers. Bring them astern."

"Aye, Captain."

The two men, legs and arms shackled, shuffled toward him, their faces alternating between deathly fear and defiance.

Columbus looked down at them. "You two have tried to get my crew to rebel against my command. I want to know why."

Silence.

Columbus nodded to an officer behind the prisoners, who struck one of them behind his knees, causing the man to scream in pain and fall onto the deck. Columbus looked at the standing mutineer and was about to nod again when the man began talking.

"If we continue on this voyage, we will all die."

"Why? Surely you are not afraid of falling off the end of Mynoa? That myth is utter rubbish. We know our world is a sphere because Tyran has smoothly descended in the east as we sail westward. If our world were flat, our mother planet would remain where it always was as we sail."

"Of course, Excellency. You have convinced us all of that. In fact, I knew it from childhood. My father had traveled to the Orient and seen Tyran in different locations in the sky. He told us all about it."

"Then why are you inciting mutiny?"

The man looked down, his long unkempt hair hiding his face.

"Look at me when I talk to you," Columbus roared. The man's head jerked upward, hair flying over his shoulders. "I will give you the count of three. If you have not begun telling me why, and telling me the truth," he said with a finger pointing at the man, "I will have you thrown overboard. One...two..."

"It was a prophecy that when Tyran sets, we enter a place where there is no air. We will all suffocate."

There was muttering behind him, some men nodding, others shaking their heads. Columbus stared at the man for a moment before breaking out laughing. "That is the most preposterous thing I have ever heard. Who told you this?"

The prisoner closed his eyes, his face contorting into an "all is lost" expression. He shook his head. "I...I cannot say, Excellency."

"And why not, pray tell?"

His lips pursed tightly, the man shook his head.

"Throw him overboard," Columbus said, turning to look at the stern and Tyran, sinking lower on the eastern horizon.

There were the sounds of a scuffle, shouting, pleading...a splash.

Columbus turned back to the other man, still on the deck. "On your feet," he hissed.

The man was up in an instant, his knees bent in pain.

"Now tell me."

The man looked back at the other crewmen pretending to lounge around the deck behind him. "I am in an awkward position, Excellency. If I don't tell you, you will have me killed. If I do tell you, the men," he paused, "will do the same when I go below."

Columbus considered. "You are indeed in an uncomfortable position. We will go to my cabin and continue this conversation." He nodded to the soldiers guarding the man, who pulled him roughly to the door of Columbus's stateroom.

There they waited until Columbus entered and, after several minutes, he commanded the man to join him. When the soldiers followed, Columbus shook his head, held up two pistols loaded and primed, and ordered them to wait outside.

When they closed the door, he strode back to his desk, sat, and addressed the man standing at attention in front of him. "If you are truthful to me, I will arrange for you to be transferred to the *Pinta*. No one there will be told why you are coming aboard, nor will the men on that ship be allowed to talk to the men here until we are in China. There, I will put you ashore. You have my word."

The man considered for a split second, then nodded. "Isabella," he whispered.

"The Queen?" Columbus demanded, astonished.

The man shook his head. "She is...uh...she is working below as a..."

"There is a woman on this ship?" Columbus roared.

"Several, in fact. Isabella is, among other things, a seer. She sees things that others cannot. She has made several prophecies already that have come true."

For several very long minutes, Columbus pondered his options. Then a plan began to develop. He explored it from every angle. Satisfied, he said, "You will be transferred immediately." He went to the door, opened it, and issued orders. Sails were furled and within fifteen minutes, a boat was pulling away from the *Santa Maria,* plowing roughly through the waves.

Columbus made no effort to round up the women. Instead he ordered an extra ration of grog for all hands. Nevertheless, he ordered the commander of the marines to double the guards on every watch, saying only that with Tyran setting in the coming week, they must be extra vigilant in case any of the other men had "doubts."

"Tyran is gone," the lookout shouted as the planet sank below the horizon under a calm sea in the mid-afternoon of October 5, 1492. A steady wind carried the three ships of the fleet westward at nearly five knots. Columbus made it a point to come out of his cabin in full dress uniform and slowly walk forward, inspecting every inch of the ship, continually breathing in deep drafts of air. The men watched him in silence. He could see them sniffing at the air and then trying deep breaths themselves.

Zonset came as never before that evening. Throughout history, people in Eurasia always saw their planet Tyran in waxing crescent phase as the Zon set. Nights never got particularly dark as the planet got fuller and fuller in the night sky. Now when the Zon set and Tyran also was below the horizon, there was no light source in the sky at all. When the orange Zon sank beneath the waves, darkness such as they had never seen before fell upon them. The fear this effect created even crept through Columbus's bravado. He looked down at his hands in the darkness, marveling at being outdoors and being unable to see them.

"Look," someone shouted. "Up there."

Eyes turned toward the sky and for the first time Europeans saw the Milky Way.

"Some of them … the stars … they have different colors," someone said. She shouldn't have, but Columbus said nothing.

"The sky is falling," someone shouted as a shooting star swept gracefully across the sky. But nothing came of the event. They sailed on in the inky darkness.

Over the next week the members of the expedition viewed countless marvels in the skies, things that had never been seen with Tyran in the sky. Throughout the long night they saw countless stars, most apparently fixed relative to each other, but a few that seemed to wander night by night: bright-tailed bodies that moved across the sky in stately arcs, and shooting stars by the dozens, some of which hissed as they flew, some of which exploded.

Equally as astonishing was the absence of an eclipse during the day. Every day of their lives before October 5, they had seen the sky darken at some time during the day as the Zon was blocked by Tyran. But not now. A full fifty-one hours of light and fifty-one hours of darkness each day.

The ship's surgeon came to see Columbus. "This new cycle of continuous light during the day and true darkness at night is causing many members of the crew to be ill. I don't really understand what is happening, but they have become morose and unable to do their jobs as well as when Tyran was 'up.'"

Columbus nodded. "To tell you the truth, I haven't felt well myself."

On the morning of October 12 a seagull landed on a yardarm of the *Santa Maria*. As many men as possible scrambled up the rigging, looking eagerly for land in the west. It was near Zonset that the cry went out, "Land ahoy." Arms pointed. They were approaching an island and if the smoke rising in several places was any indication, it was inhabited.

"China," Columbus said, smiling at the officers around him. He summoned the commander of the marines.

"Our arrival is complicated by the fact that we must leave certain people from the crew on that island." He looked at the officer carefully. "As you know, there are women on this ship and, I suspect, the others. You will arrange for them and the sailor Diego whom I sent to the *Pinta* to be discreetly left there at the end of our visit. Is that clear?"

The officer nodded. "And their children, too, sir?"

Since time immemorial, our ancestors tried to make sense of our Moon and its alien surface. Although its colors differ from those of the Earth, they saw patterns on it that suggested that some of the Moon's features might be analogous to things found on Earth. To them, and indeed, to us, the roundish, dark regions of the Moon appear to be seas, surrounded by lighter regions that are dry land, with mountains and valleys. Human imagination being what it is, these similarities encouraged storytellers in many cultures to populate the Moon with people in a quasi-Earthlike setting.

Science eventually put an end to that speculation. Just over 2,000 years ago measurements, starting with those of Aristarchus of Samos, revealed that the Moon is much smaller than the Earth. In the nineteenth century, telescopes showed it to be a waterless, airless world. In the twentieth century, technology was able to transport us there, reaffirming the Moon's inability to sustain life.

We are used to viewing our Moon from on high, namely from the larger Earth. However, the size relationship between the inhabited Earth and uninhabited Moon could well have been reversed. This is what we explore in this chapter: what if an Earth-sized, inhabited moon orbited a larger uninhabitable planet? I call the Earthlike moon Mynoa and the larger planet it orbits Tyran. Based on the masses of the giant planets in our solar system—Uranus, Neptune, Saturn, and Jupiter—Tyran's mass could be between roughly 10 and 325 times the mass of the Earth. I choose Tyran to have the same mass as Neptune, 17 times that of Earth. Tyran is to be a clone of Neptune.

Placement of this chapter's pair of worlds in orbit around their star, the Zon, is restricted by the condition that the water on the inhabited moon's surface must be able to be liquid. Liquid surface water is essential for the formation and early evolution of life as we know it. It is in that water that organic molecules form and interact myriad times, leading to the formation of living cells and then more complex life. Although Tyran is identical to Neptune, Tyran and Mynoa can't orbit out where Neptune does, thirty times farther from the Sun than the Earth. At Neptune's distance from the Sun, where the air temperature would be −360°F, surface water on Mynoa would freeze as solid as concrete. The region of a solar system where the surfaces of worlds are inhabitable is called the Habitable Zone. Simply put, it is the range of distances from the star at which water can be liquid on the surfaces of the planets or moons.

The inner edge of the Habitable Zone around our Sun lies between the orbit of Venus (which is 0.72 times as far from the Sun as we are) and the orbit of the Earth. Closer to the Sun than this boundary, the water on a planet or moon is vaporized, which is why the surface of Venus is bone dry. The outer edge of the Habitable Zone is located just inside the orbit of Mars (which is 1.52 times as far from the Sun as we are). To make matters as simple as possible, I therefore chose to put Mynoa and Tyran in the same orbit as the Earth.

This sounds plausible enough, perhaps even the obvious place to put them, but until the mid-1990s, I would not have been scientifically justified in making this choice, given that Tyran is Neptunelike. Based on the fact that all the giant planets in our solar system are much farther from the Sun than the Earth, most astronomers before that time believed that giant planets just couldn't exist as close as we are to the Sun. We were wrong.

BUILDING A SYSTEM OF PLANETS

Basically two families of planets form in the disk of gas and dust surrounding a newly forming star (see the discussion in Chapter 1). Those like the Earth, called terrestrial planets, are composed primarily of rock and metal. Giant or Jovian planets, on the other hand, are composed of liquid hydrogen, helium, water, methane, ammonia, and a terrestrial core thrown in for good measure. The terrestrial planets in our solar system are Mercury, Venus, Earth, and Mars. Although Earth is the largest of them, it is much smaller and less massive than any of the giant planets, of which Jupiter is the largest and most massive. All the terrestrial planets in our solar system orbit relatively close to the Sun compared to the paths of the giants.

Astronomers in the olden days, prior to 1996, thought that all giant planets must exist out beyond the orbit of Mars. When a disk of gas and dust forms around a young star, all the elements and molecules are distributed randomly through it. The theory was that the inner region of the disk quickly becomes too hot for the light gases, such as hydrogen, helium, methane, and ammonia to remain there and form planets. Heated, these gases were believed to drift outward, away from the star, before any planets formed. Out where the giant planets orbit in our solar system, however, the temperatures are so low that the light gases were able to stay there and become parts of those planets.

Traditionally, terrestrial planets, including some out where giant planets are today, were believed to have formed before the giant planets. Terrestrial planets are composed of dense elements such as silicon, carbon, iron, and nickel, which remain scattered throughout the disks around stars long enough for planets to form from them. This orbiting debris collides, clumps together, and eventually merges into a few large blobs called protoplanets that grow to become planets. Once the terrestrial planets had formed in the outer reaches of our solar system, the pre-1996 planet-formation model predicted, the light gases then fell onto them in such large quantities that these worlds were transformed into giant planets.

This scenario is still viable in explaining how our solar system came to have the planets it has, but other circumstances have been proposed to explain the hundreds of giant planets that have been discovered, many of them closer to their stars than our planet Mercury is to the Sun! One theory proposes that all giant planets still form far from their stars, as just described, but then somehow some of them migrate inward. Another theory proposes that unlike the giant planets in the solar system, which took tens or hundreds of millions of years to coalesce, giant extrasolar planets especially close to stars formed very rapidly, before the lightweight gases there could be blown away.

These theories can be tested because star formation occurs in many places throughout our Milky Way Galaxy today. You can see one such region in the night sky when the constellation Orion (The Hunter) is visible. First, locate Orion's belt. Below it "dangle" three stars called the sword of Orion. Look at the middle "star" of the sword and you will notice, even with your naked eye, that it looks peculiar, too irregular to be a star. In fact, that is the central region of a vast, star-forming, interstellar cloud called the Orion molecular cloud, located about 1,500 light-years from Earth and containing enough gas and dust to form some 200,000 stars. Astronomers study such star-forming regions to see if giant planets are forming close to young stars in them.

A light-year, the distance light travels in a year, is about 5.9×10^{12} (5.9 trillion) miles, meaning that the Orion molecular cloud is about 8.9×10^{15} miles away. To give you an idea of how far away this is, traveling at 80,000 mph, typical of spacecraft leaving the solar system today, it would take you about thirteen million years to get to the Orion molecular cloud from here. For convenience, our distance from the Sun, a mere 93×10^6 (93 million) miles, is designated by astronomers as 1 astronomical unit (AU). Mercury is .3 AU from the Sun. The closest giant planet to the Sun, Jupiter, is 5.2 AU from it. In some star systems, the giant planets closest to their stars are less than 0.05 AU from them. The daytime surface temperatures on these giant planets are over 3,800°F, which is hot enough to melt nickel, silicon, iron, and dozens of other substances here on Earth.

Planets orbiting other stars are called extrasolar planets. The discovery of giant extrasolar planets close to their stars justifies my asking the scientific question, "What if a Neptunelike planet had formed here at Earth's present distance from the Sun and an Earthlike moon had formed in orbit around it?" As noted above, I call the Earthlike body Mynoa and its mother planet Tyran. They orbit a star called the Zon, which is identical to the Sun.

MYNOA-TYRAN

In contrast to our solar system, the Zon's formation was followed very quickly (compared to the formation of our Earth) by the condensation of the massive planet Tyran. As you can see, I am assuming that the model of rapid formation of a giant planet is correct. Tyran condensed from vast quantities of hydrogen, helium, water, ammonia, methane, rock, and metal orbiting the young Zon. Tyran has a core of rock and metal, surrounded by an ocean of water thousands of miles deep, which is surrounded in turn by a shell composed primarily of liquid hydrogen and helium about 2,000 miles thick. This outer layer tapers outward into an atmosphere.

We are now ready to flesh out the new worlds of this chapter with details of their physical properties.

- Tyran has 17 times the mass and 3.9 times the diameter of the Earth.
- The plane in which Tyran orbits around its star is called the ecliptic, which is also the name assigned to the plane of Earth's orbit around the Sun (see Figure 1.2).
- Tyran's axis of rotation (or spin) is perpendicular to the plane of its orbit around the Zon (vertical lines in Figure 2.1).
- Mynoa orbits above Tyran's equator, which is also the plane of the ecliptic (see Figure 2.1). You can also see from the figure that the Zon is always directly above Mynoa's equator.

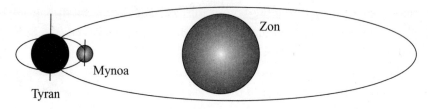

Figure 2.1: *Orbits and rotation axes of Tyran and Mynoa. The bodies in this figure are not drawn to scale.*

We show shortly that all these orbital properties play important roles in the characteristics of life on Mynoa.

Formation of Mynoa

As with Lluna and Kuu in Chapter 1, we need to choose a method of getting Mynoa into orbit around Tyran. One possibility for the formation of Mynoa is that it formed far from Tyran and then was captured by it, like Lluna in the previous chapter. This is unlikely in the extreme because Mynoa, with $1/17$ the mass of Tyran, is so massive that it would be virtually impossible to capture even if Mynoa had a moon to shed.*

Another possibility is that a large body smashed into Tyran and splashed debris into orbit around it as with the formation of our own Moon. However, Tyran's outer liquid layer, mostly hydrogen and helium, would vaporize if struck by a giant impacting body. These gases would leave the vicinity of Tyran forever, rather than stay in orbit and form rocky Mynoa.

The possibility of Mynoa spinning off Tyran (see the Appendix) must also be excluded for the same reason, namely the outer layers that would be shed are the wrong chemical composition.

The final possibility is that Tyran itself could have had a ring of debris orbiting it when it first formed, like a microcosm of the disk orbiting the Zon from which Tyran and the other planets there were formed. That debris could then have clumped together and become Mynoa. Saturn's four largest

---

* Lluna in Chapter 1 had $1/81$ the mass of Dimaan. Neptune's moon Triton, which almost certainly was captured by the planet, is relatively tiny compared to the bodies we are discussing, with only $1/5,000$ the mass of Neptune.

moons are likely to have formed this way. Allowing that chaos could create a single massive moon around Tyran instead of several smaller ones, I chose to form Mynoa this way.

Properties of Giant Tyran

Let's briefly tour the young planet Tyran. I give it rotation (spin on its axis) so that the Zon goes from one high noon to the next in twelve hours. That means the day on Tyran is twelve hours long, which is similar to Neptune's present sixteen-hour day.* There is a subtlety in specifying the length of Tyran's day: it doesn't have any surface from which to watch the Zon rise and set. I don't just mean that it doesn't have a solid surface. It doesn't have a liquid one either! Indeed, neither Neptune nor Tyran has a surface on which you could land or into which you could splash. Let me explain.

Here on Earth, you know you have landed in a swimming pool after you jump off a diving board, especially if you do a belly flop. On Tyran or Neptune, however, the atmospheres of hydrogen and helium get denser and denser as you descend through them, but this region lacks a boundary that you would call the surface, as you would find between the water in your pool and the air. On the giant planets, you eventually get into sufficiently dense hydrogen and helium to say, "Hey, this is liquid," but getting there does not involve crossing a boundary. Technically, we say that there is no "phase transition" in going between gas and liquid on the surfaces of these planets. Although the outer layers do rotate, the length of the day for both Tyran and Neptune is defined by the time it takes the terrestrial planets deep in their cores to rotate once. That is, if you were standing on the core and could see through the thousands of miles of liquid above you, you would see the Zon go from noon to noon in one "day."

Location, Location, Location

One of the foundations of planet- and moon-building is location, as we have already seen in deciding where to have Tyran and Mynoa orbit the

---

* A twelve-hour day isn't outrageous, considering that Jupiter, which has nearly three times the diameter of Neptune, has a ten-hour day.

Zon. Choosing a suitable orbit for Mynoa around Tyran is a little tricky. In the first place, Mynoa needs to be a substantial moon, held together by its own gravity, rather than a big rock orbiting Tyran. If Mynoa were a small, potato-shaped moon, it would lack sufficient mass and hence sufficient gravity to hold an atmosphere or oceans, so it would not be suitable for life. No potatoes for you. Two distances from a planet need to be considered in locating any substantial moon in orbit around it: the Roche limit and the synchronous orbit radius.

The Roche Limit

Orbiting debris must be a minimum distance from a planet in order for a substantial moon to develop from it. These moons have diameters of 200 miles or more, and are essentially spherical in shape. This minimum distance is called the Roche limit after Édouard Albert Roche (1820–1883), who first calculated it. Inside that distance the planet's gravitational force pulls matter apart faster than it can combine. In other words, as clumps of debris orbiting inside a planet's Roche limit drift near each other, the gravitational force from the planet pulls them apart, preventing a substantial moon from coalescing there. Because Saturn's rings are located inside that planet's Roche limit, they will never become a moon. Indeed, no substantial moon forms inside the Roche limit of any planet. Tyran's Roche limit is about 2,000 miles above its surface.

Synchronous Orbit

The other crucial distance beyond which substantial moons must form is the radius of synchronous orbit (called geosynchronous orbit for Earth). This is the distance at which a moon orbits at the same rate that the planet below it spins on its axis. Closer to the planet, the moon orbits faster than the planet spins. Beyond this distance the moon orbits more slowly than the planet rotates. Because they are going around at the same rate, a moon in synchronous orbit above a planet's equator appears to hover in one place in the sky over the planet. We put satellites in geosynchronous orbit around the Earth. When they are fixed over our planet, we can continually bounce signals off them.

The reason moons must form at or beyond synchronous orbit is related to tides. In the first chapter we saw that a moon like ours, orbiting outside synchronous orbit, creates tides on the planet. The closer tide pulls it forward in its orbit, thereby giving the moon energy to spiral outward. Our Moon, for example, is receding from Earth at a present rate of about an inch and a half per year. Conversely, a moon inside synchronous orbit creates tides on the planet that pull it backward, causing it to lose energy and spiral inward. Eventually moons in these latter orbits would spiral down to the Roche limit and be pulled apart, forming a ring around the planet. Since we want Mynoa to persist so that life can evolve on it, I have that moon form beyond synchronous orbit.

The distance between a moon in synchronous orbit and its planet depends on how fast the planet is spinning. That distance can be determined from the work of the German mathematician and astronomer Johannes Kepler. In 1605, using the observational data of his mentor Tycho Brahe (aka Tyge Ottesen Brahe), Kepler deduced an equation describing the relationship between the period of a planet's orbit around the Sun and the distance between the two bodies. In general, the farther the planet is from the Sun, the longer its orbit. His equation was generalized to work for moons orbiting planets by Sir Isaac Newton (1643–1720).*

As our Moon and Mynoa spiral away from their respective worlds, the friction generated by the tides in the oceans of Earth and Tyran cause these two worlds to rotate ever more slowly. Indeed, our day is about .001 seconds longer than it was a century ago. Although this seems like a small change, geologists have evidence that the day here after the impact creating the Moon occurred was between five and six hours long. Through the eons, friction between the oceans and land has slowed that rotation to a twenty-four-hour day.

Contrary to what seems obvious here, neither our Moon nor Mynoa is ever going to leave its home world entirely. As the Moon and Mynoa spiral outward and their respective planets rotate more and more slowly, eventually

---

* If it were in synchronous orbit, Mynoa would be fixed over one side of Tyran and hence never visible from the other side of the planet. Using Newton's equation for the period of orbits, we can calculate the distance from Tyran at which Mynoa is in synchronous orbit. When young Tyran had a twelve-hour day, the synchronous orbit around it was 28,000 miles above its surface (43,000 miles from its center).

the Earth and Tyran will be rotating at the same rates that the Moon and Mynoa are orbiting them. In other words, even though moons are unlikely to start in synchronous orbits, the combination of a moon moving outward and its planet rotating more slowly will eventually force the smaller bodies into synchronous orbits. Thereafter, the Moon and Mynoa will remain at that distance from their parent worlds. It is worth noting that the Sun and the Zon are going to explode long before this takes place.

Let's focus now on Mynoa's orbit. We know that it must form beyond Tyran's synchronous orbit radius. Guided by this distance, I have chosen to create Mynoa nearly 30,000 miles above Tyran's surface, which is a couple of thousand miles farther than the synchronous orbit. Put another way, Mynoa will form 45,000 miles from the center of Tyran. As measured with respect to the distant stars, it will take Mynoa 12.7 hours to orbit the larger body. From what we have just discussed we know that Mynoa will immediately begin creating tides on Tyran and, as a result, spiral outward.

The period of a moon's orbit around its planet depends on the planet's mass, the moon's mass, and the distance between the centers of the two bodies. Clearly, the only thing that changes substantially over time in our scenario is the distance between Tyran and Mynoa. As the latter spirals outward, its period of orbit increases. For example, our Moon took only 10 days to orbit the Earth when it was half its present distance from us, whereas now it takes about $27\frac{1}{3}$ days to go around once. Between the time Mynoa forms and the time it quadruples that distance from the center of Tyran, the period of Mynoa's orbit will change from 12.7 hours to 102 hours.

Mynoa's Rotation

The initial orbit around Tyran that I have chosen for Mynoa has repercussions on Mynoa's rate of rotation. Like the young Earth, I set Mynoa to initially rotate once every eight hours. Similarly, when Mynoa formed, it had a molten surface, heated by the decay of radioactive elements inside it and by the unremitting impacts of space debris its gravity pulled down onto it. Mynoa created huge tides of liquid hydrogen and helium on Tyran, and the larger body generated enormous tides of molten rock on Mynoa. As on Tyran, Mynoa had two high tides, one on the same side as the larger body and the other on the opposite side from Tyran.

Mynoa's early rotation dragged its molten rock tides around with it. As a result, the two high tides were not aligned between its center and the center of Tyran. This should sound familiar, as it also applied to the tides on Tyran and to the worlds of Chapter 1. If we could hover safely over young Mynoa, we would see colossal tides of molten rock thousands of times higher than the water tides our Moon creates on Earth today. It's hard to imagine the violent heaving tides of molten rock that rose and fell two hundred miles or more every few hours on young Mynoa.

As with the evolution of Lluna in Chapter 1, the friction between the molten rock tides and the rest of that body will cause the rotation rate to slow until Mynoa is spinning on its axis at exactly the same rate that it is orbiting Tyran. At that point, the land tides will stop flowing over the surface of Mynoa; one will be fixed directly below Tyran, whereas the other will be on the opposite side of Mynoa from it. With the end of the tidal flow on its surface and the loss of the heat that the moving tides have generated, the surface of Mynoa will cool enough to form a solid crust. Keeping one high tide directly between the centers of Mynoa and Tyran requires that Mynoa rotate at the same rate as it revolves around the larger body. In other words, Mynoa will have slowed into synchronous rotation (introduced in Chapter 1) around Tyran.

Once Mynoa is in synchronous rotation around Tyran, half of Mynoa never sees Tyran, and Tyran remains fixed in the sky as seen from everywhere on the other half of Mynoa. For convenience, let's call the side of Mynoa forever facing Tyran the near side and the side facing away from the larger body, the far side.

Despite its synchronous rotation around Tyran, the changing land tides do not vanish completely on Mynoa. The remaining ones are caused by the Zon. As Mynoa goes around Tyran, the Zon creates minuscule land tides amounting to about one-third of what we experience today on Earth. Because the highest land tides on Earth are about 8 inches, and one-third of their height comes from the Sun, the highest land tides on older Mynoa will be one-third of 8 inches or about 3 inches high.

Length of the Day on Mynoa

The length of the day profoundly affects life on any world. A day, formally a solar day, is the time from one noon (when the Sun is highest in the sky) to the next. The day is determined by how fast a moon or planet rotates, combined with how fast it orbits its star. A body's rotation causes the Sun or the Zon, along with everything else in the sky, to rise and set, hence creating the effect we experience as day and night.

The length of the day is the primary factor determining our daily activities, such as waking and sleeping, eating, hormone secretion, blood pressure, brain wave activity, physical coordination, alertness, and body temperature, among others. Indeed, most life-forms on Earth are regulated by the twenty-four-hour cycle in which we live. Because day length is so important to life, we need to determine the number of hours in the new world's day for each version of the Earth in this book. We have already seen that Earth's day was originally about eight hours long, decreasing to about five hours after the impact, and lengthening since then to its present twenty-four-hour period.

The length of Mynoa's day will change even more dramatically. Like Earth, Mynoa's day was originally 8 hours long. At that time, Mynoa orbited around Tyran once every 12.7 hours. As Mynoa spiraled outward, it was transformed into synchronous rotation by the molten rock tides on its surface. In other words, its rotation slowed from once every eight hours to once every orbit around Tyran. Because the time it takes a moon or planet to rotate determines its day, as Mynoa's rotation rate slowed, its day lengthened.

As Mynoa continues to spiral outward, the time it takes to orbit Tyran increases, as shown by the equations of Kepler and Newton. Because the time it takes Mynoa to orbit its planet is also Mynoa's rotation period, the increase in the length of the orbit causes the days to lengthen. Five and a half billion years after it formed, Mynoa had spiraled out to a distance of 180,000 miles* from Tyran. The orbital period of Mynoa, and hence the day, would be roughly 102 hours long; this is the distance from Tyran at which Mynoa will be when sentient life evolves.

Formation of the Oceans and Water Tides

We are now ready to beginning "dressing" Mynoa. Like its exemplar, Earth, the first thing Mynoa will wear is water, which existed as ice in the

---

* This number is not cast in concrete due to the complexity of the tides created on Tyran by Mynoa, but serves our purposes well.

cloud of gas and dust from which Zon and its host of orbiting bodies formed. At first, much of that water clumped together with rocky debris to create comets. Because of their composition, comets are often called dirty icebergs in space.

When Mynoa's surface solidified, water began to accumulate on it. Much of that water derived from comet impacts, as is true of our Earth, along with moisture freed from the interior that did not evaporate into space. The evidence we have for comet impacts providing water comes, among other places, from the four large moons of Jupiter. Called the Galilean moons, virtually all the impact craters on them were created by comets. In our scenario, similarly, many comets passing Tyran would be pulled around that planet and strike Mynoa.

Once locked into synchronous rotation around Tyran, Mynoa's oceans have rising and falling tides that are always much lower than the tides we experience on Earth. When oceans formed on the young Earth, towering water tides were generated primarily by the Moon (which was much closer then than it is now). Because the Earth's rotation was not synchronous with the orbit of our Moon, ocean tides swept around our planet (as they still do today). As the Moon spiraled away, the ocean tides decreased in height to what they are now. Like the Earth, Mynoa developed a pair of high ocean tides, generated by Tyran, but because Mynoa was by then in synchronous rotation around Tyran, those tides were immediately fixed in the same places as the high land tides. That is, they created one fixed, high ocean tide directly under Tyran and another on exactly the opposite of the smaller body.

Complicating matters, Mynoa also has a small, moving, ocean tide created by the Zon. That tide, which does orbit Mynoa, is one-third as high as the tides we have on Earth today. The period of these moving ocean tides (from high tide to low and back to high again) is different than it is for the Earth. The length of Earth's cycle of tides is determined by the location of our Moon in the sky, because the Moon is the dominant cause of tides here. Recall from Chapter 1 that two high tides sweep across opposite sides of the Earth: one where the Moon is high in the sky and the other on the opposite side of the Earth. These tides are continually on the move. Because there are two high tides passing the same place each day, an interval of 12 hours and 26 minutes occurs between one high tide and the next at the same place. The extra 26 minutes is a result of the Moon moving around the Earth as the tides flow.

Because Mynoa has no moon and because Tyran is fixed in Mynoa's sky, the period of the moving tides on Mynoa is determined by the location of the Zon. In particular, the cycle of tides takes half the length of a Mynoan day. When Mynoa and its oceans are young, the cycle of water tides would be ten hours (half a twenty-hour Mynoan day), whereas older Mynoa will have an ocean tide cycle of fifty-one hours.

Tyran as Seen from Mynoa

We are used to our Moon being the largest object in the night sky, but it pales in comparison to the view of Tyran from Mynoa. Consider, first, the evolution of our Moon's appearance from Earth. When the Moon formed, it was some ten times closer than it is today. At that time, it would have appeared ten times larger in angle than it does today. Originally, then, the Moon spanned five degrees in the sky, which is about the size of your fist held at arm's length. Today the Moon fills only half a degree in our sky. In comparison, when Mynoa formed, 30,000 miles above the surface of its planet, Tyran spanned a staggering thirty-seven degrees. That is, it spanned over one-fifth of the sky. As Mynoa spiraled away from it, Tyran appeared smaller and smaller. With Tyran 180,000 miles from Mynoa's surface, Tyran still spans an impressive nine degrees, or eighteen times the angular size of our Moon as seen from Earth today.

Seasons on Mynoa

Mynoa will have very, very little seasonal change in temperature. It's not that I dislike seasons. I wouldn't live in Maine if I did. Mynoa lacks seasons because the event that would have caused them, a gigantic impact that would tilt Mynoa's rotation axis over, never happened there. As we saw in Chapter 1, most astronomers believe that a Mars-sized body wandering through our young solar system struck the Earth. Not only did that impact splash debris into orbit, it also knocked Earth's rotation axis over about $23\frac{1}{2}°$ (Figure 2.2). Mynoa was never hit, so it wasn't knocked over.

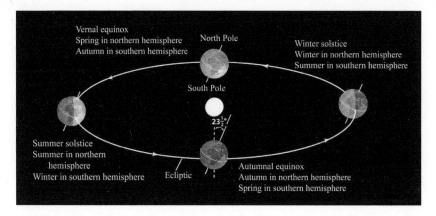

Figure 2.2: *Earth's seasons are caused by the tilt of the Earth's axis.* CREDIT: W.H. FREEMAN & CO.

It's interesting to understand the seasons on Earth and, hence, the lack of them on Mynoa. Seasons are caused by the tilt of the rotation axis and by the ellipticity (oval shape) of a planet's orbit around the Sun. Because of the tilt of the Earth's axis relative to the ecliptic, the Sun actually rises in different places on the eastern horizon on different days of the year, moves to different maximum heights each day, and remains "up" for different lengths of time throughout the year. The summer is the warmest season because the Sun rises highest in our sky and is up the longest amount each day during that time of the year. Conversely, winters are coolest because the Sun rises later, does not rise nearly as high in the sky, and sets earlier than during any other time of the year. The height in the sky is relevant because the higher the Sun goes the more intense or concentrated is its energy on Earth.

In contrast, Mynoa's rotation axis is at right angles to the ecliptic (see Figure 2.1), so that at any place on that world, the Zon rises and sets at the same location and at the same time every day of the year. The amount of heating that the Zon provides to a given place under those circumstances is the same day in and day out. By the way, most astronomers believe that the Earth was formed spinning with its axis perpendicular to the ecliptic, the same way I have set Mynoa's axis, which means that Earth lacked seasons back then.

As noted above, the other factor contributing to the seasons is the shape of the orbit around the Sun or Zon. Neither the Earth nor the Tyran–Mynoa system has a circular orbit. We are closer to the Sun by nearly four million miles in January than in July. That difference in distance has very little effect on Earth because our northern and southern hemispheres have different amounts of land and water from each other: when we are closest to it, the Sun is high in the sky over our southern hemisphere, which is mostly covered with oceans. They send more heat right back into space than the land does, so the Earth remains cooler during that time than it would be if the southern hemisphere had more continents. When we are farthest from the Sun, it is highest in the northern sky, heating the continents more than it heated the southern oceans six months before. The changing distance and the changing amount of heat the Earth retains just about cancel each other out, so our seasons are affected only by a few degrees due to our changing distance from the Sun.

Because Mynoa's axis is not tilted, the only cause of seasons will be the tiny effect of the changing distance from the Zon. When Tyran and Mynoa are closer to the Zon, the temperature will be several degrees warmer everywhere on Mynoa than when the two bodies are farthest from it. Unlike the Earth, this means that every place on Mynoa feels the same slight temperature change throughout the year, rather than the northern hemisphere being hotter while the southern hemisphere is cooler, and vice versa.

Phases of Tyran

Tyran, unmoving in the Mynoan sky, still goes through a cycle of phases as seen from the smaller world. Although Tyran's lack of apparent motion differs from our Moon's daily motion across the sky, the phases of Tyran as seen from Mynoa and the phases of our Moon as seen from Earth are the same.

The changing phases of Tyran, which are the changing fraction of the Zon-lit side seen from Mynoa, occur because Mynoa is orbiting around that planet, just the opposite of the relationship we have to the Moon. When Mynoa is between the Zon and Tyran, Tyran appears full. When Mynoa is on the opposite side of Tyran from the Zon, Tyran appears new. The entire cycle of phases seen from Mynoa is much shorter than the 29½-day cycle of lunar phases seen from Earth. This occurs because Tyran goes through a cycle of phases every day on Mynoa. When Mynoa was young, the cycle of

Tyran's phases seen from Mynoa's surface was only about 12.7 hours, lengthening in time as Mynoa spiraled away from Tyran. When advanced life evolves on it, Mynoa's orbit around Tyran takes 102 hours, so the cycle of phases seen from Mynoa at that time will be only 102 hours long (equivalent to just 4¼ Earth days).

ECLIPSES

Eclipse Shadow on Tyran

Eclipses on Earth are events of passing interest, but they are so important on Mynoa that they affect the evolution of life there. Tyran makes such a large angle in Mynoa's sky that there will be eclipses during every new and full Tyran. When Tyran is full, Mynoa's shadow is assured of striking it. The shadow won't cover all of Tyran, however, as a result of Mynoa's relatively small diameter and large distance from its planet. This is analogus to the Moon's shadow on the Earth. The eclipse of Tyran by Mynoa will last about 35 minutes when the day on Mynoa is 12.7 hours and increase to 1 hour and 40 minutes when they are four times farther apart. As noted earlier, it is at this latter distance, some 5½ billion years after it formed, that I posit advanced life evolving on Mynoa.

Eclipse Shadow on Mynoa—Zolar Eclipses

The Moon's shadow on Earth creating a solar eclipse only cuts through a narrow band on our planet, never wider than about 160 miles. Thus, you have to be at certain locations here to see a total solar eclipse, one in which the Sun is completely blocked. Our tourist industry does a brisk trade transporting people to sites where solar eclipses can be seen. Even if you fork over thousands of dollars to see such an eclipse, the Moon's shadow moves so fast across the Earth's surface that total solar eclipses last for a maximum of only about seven and a half minutes. Will the tourist industry give more bang for the buck on Mynoa?

The moon Mynoa moves into the shadow of Tyran in that planet's new phase. Tyran's shadow is so wide that during these events all of Mynoa is eclipsed by it.* From the perspective of someone on Mynoa, these Zolar eclipses are the equivalent of solar eclipses on Earth. Weather permitting, everyone on Mynoa's near side can see each Zolar eclipse. The travel industry won't make a penny taking people on Mynoa's near side to see eclipses. Of course, no one on the far side of Mynoa will ever see a Zolar eclipse from there, because they never see Tyran. The travel industry could make a fortune taking far-siders to the near side to see eclipses.

The time of day on the near side during which a Zolar eclipse occurs depends on where Tyran is in your sky. If you live at a place where Tyran is fixed near the eastern horizon, the daily Zolar eclipse will occur in the morning. If Tyran is high in your sky, the eclipse occurs around midday. If it is in the west, the eclipse occurs in the afternoon. A Zolar eclipse occurs every day everywhere on Mynoa's near side.

Day and Night on Earth

Most people on Earth are used to the Sun rising at some time in the morning, followed by it moving higher in the sky, reaching a maximum height, then moving lower in the sky, with the hours of daylight ending at sunset. On two regions of the Earth, however, the cycle is different. They occur above latitude 66½° north (the Arctic Circle) and below the equivalent latitude in the south (the Antarctic Circle). In these realms, the Sun rises and sets "normally" for part of the year. However, for days, weeks, or months during winter (depending on the latitude) the Sun never rises there at all. Conversely, for days, weeks, or months during the summer, the Sun never sets. The realms of the "midnight Sun" are determined by the 23½° tilt of the Earth's axis.

Day and Night on Mynoa's Far Side

Let's see what the day is like on Mynoa when people come to exist there. First, consider a day on its far side, where Tyran is never visible. Every cloud-free night of the year the sky blazes with the light of stars that are never dimmed by Tyran, unlike on Earth where stars as seen from everywhere are

---

* In fact, Tyran's diameter is as much larger than Mynoa's as Earth's diameter is compared to that of our Moon.

often dimmed by the light of our Moon. This problem (from the point of view of us astronomers on Earth) is most pronounced during the gibbous and full phases of the Moon.

Time passes and the Zon peeks over the eastern horizon: it is Zonrise. The Zon ascends to its maximum height in the sky and then it sets, just like in your hometown. Somewhat subtle variations exist between the days experienced on Earth and those on Mynoa's far side related to the different tilts of the two planets (see Figures 2.1 and 2.2). Recall that Mynoa's rotation axis is perpendicular to the ecliptic. As seen from any given place there, this orientation of the axis causes the Zon to rise straight up and set straight down in the same places every day of the year.* Also, the number of daylight hours equals the number of nighttime hours throughout the year everywhere on this half of that world. In contrast, you know from experience that we have more hours of daylight in the summer than in the winter, that the Sun rises at an angle (except as seen at the equator), and that the Sun rises in different places on the eastern horizon and sets at different places on the western horizon on different days throughout the year.

Day and Night and Day and Day on the Near Side

The big difference in day–night cycles compared to Earth is found on Mynoa's near side. Depending on your location on that side, Tyran could be fixed in your sky anywhere from on the horizon to directly overhead. As an example of a day on the near side, let's start at Zonrise and assume that you live where Tyran is directly overhead. During the morning hours, the Zon comes up in the eastern sky and, sometime before noon, it is eclipsed by Tyran. The sky becomes nearly black,† with some stars visible. For two hours in the middle of the day, there is night. After the eclipse ends, bright daylight returns until Zonset.

It is very important to consider the phases of Tyran that you would see from this place on Mynoa throughout each day. At Zonrise, Tyran is in the third quarter phase, meaning that you would see half of it lit up. Then it

---

* Those places on the eastern and western horizons depend on your latitude on Mynoa.
† The hydrogen-rich atmosphere of Tyran will bring some Zonlight around to your side on Mynoa, so it will never be truly as dark as is normal nighttime on Earth.

becomes a waning crescent and finally moves into the new phase during the eclipse. After the midday eclipse is over, the phases of Tyran continue from new to waxing crescent to first quarter phase around sunset. Immediately after sunset, the sky, lit by Tyran, begins to brighten again. Continuing through the waxing gibbous phase, another kicker comes at midnight when Tyran is in the full phase. At that time Tyran is 2,800 times brighter than our Moon ever gets! As a consequence, midnight on the near side of Mynoa is brighter than noontime there and almost as bright as noontime here on Earth! Between midnight and sunrise, Tyran goes from full through waning gibbous to third quarter. In other words, past midnight it and the sky darken somewhat, but not enough to see more than a few of the brightest stars. Amazingly, on the near side of Mynoa three periods of bright daylight and one of relative darkness (the eclipse) occur each day.* Figures 2.3a and b show the profoundly different brightnesses throughout the day on each side of Mynoa. Figure 2.3b also shows the brightness for a day on Earth.

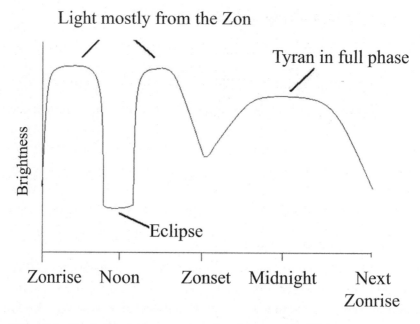

Figure 2.3a: Sketch of the brightness at the surface of the near side of Mynoa.

---

* The same story holds for every place on the near side, with the eclipse occurring at different times of the day, depending on how high Tyran is in the sky.

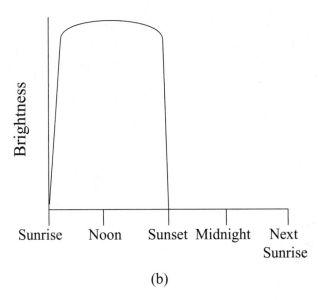

(b)

Figure 2.3b: The brightness at the surface of the far side of Mynoa and the brightness at the surface of the Earth at mid-latitudes.

The daily Zolar eclipses caused by Tyran on Mynoa have significant effects on the moon's weather. Every time the Zon goes down or is eclipsed, the air begins to cool off. Therefore, Mynoa's near side doesn't heat up as much as does its far side, where eclipses never occur. The lower maximum temperature and the three light–dark cycles per day on the near side compared to the single cycle per day on the far side cause Mynoa to develop two sets of life-forms, each adapted to the light and dark cycle of one side of that world or the other. Creating intelligent life is the ultimate goal, so let's take this in a few steps, starting with a breathable atmosphere.

Evolution of Mynoa's Atmosphere

Neither Earth's nor Mynoa's first atmospheres resembled the air we breathe. Both worlds began with atmospheres rich in hydrogen and helium, gases left over from the planet-forming and moon-forming era. The gravitational attraction of these worlds was too weak to hold these gases near their surfaces. As a result, as the hydrogen and helium were heated by the Sun, they rose upward and evaporated into space, never to return.

The second atmospheres that Earth and Mynoa developed were like the one Venus has today, composed primarily of carbon dioxide, with a small amount of nitrogen and traces of other substances including water, carbon monoxide, and various sulfur compounds. Most of this gas came from inside Mynoa, Earth, and Venus, entering the air through volcanoes and cracks in the worlds' surfaces. We still see these latter features on Earth today in, for example, the mid-Atlantic rift, which looks like an underwater mountain range that is hemorrhaging gas and molten rock.

The amount of carbon dioxide available to be released into a planet's atmosphere is quite staggering. Both the Earth and Venus began with about ninety times as much carbon dioxide in their atmospheres as the total amount of gas in the air we breathe today. This is to be the initial amount of carbon dioxide–dominated air around Mynoa.

Because gravity pulls the gases in an atmosphere downward, that gas exerts a pressure on everything it encounters.* The air we breathe today exerts 14.7 pounds of force on every square inch of your body. As with Venus today, Mynoa's early carbon dioxide–rich atmosphere contained so much gas that the air pressure on the surface of young Mynoa was what you would feel today if you went swimming two-thirds of a mile underwater in the Earth's oceans. This is an unsuitable atmosphere for life as we know it, but because that was also Earth's early atmosphere, there must be a way to morph it into a breathable one.

Earth transformed its incredibly dense atmosphere into the one we breathe in basically two steps. First, as the oceans formed and water evaporated into the air, some of the carbon dioxide in the atmosphere dissolved in that water vapor and was removed from the air as the water came down as rain. Calculations for Earth suggest that as much as half the early carbon dioxide was taken into the oceans that way. You already have an intuitive feel for how well carbon dioxide dissolves in water: the "carbonation" in soda and beer is carbon dioxide, and as you know from experience, these drinks fizz for a long time after the container is opened and the gas released.

Much of the carbon dioxide deposited in Earth's oceans was eventually taken up by early aquatic plant and animal life, whose existence we discuss shortly. Biological processes in some of these creatures combined the carbon

---

* Pressure is a force acting on an area. Pressing your palms together creates a pressure between them.

dioxide with other molecules to create their shells, among other things. As these creatures died, these shells sank to the ocean bottoms. When the layers of shells became thick enough, they were compressed into rock, such as limestone, thus locking their carbon dioxide out of the atmosphere.

Plant life helped remove the remaining carbon dioxide. At first these were plants in the oceans, taking up some of the dissolved carbon dioxide. Eventually plant life evolved onto Earth's surface and the process of removing the carbon dioxide accelerated. Carbon dioxide is essential to the survival of plant life. Plants use carbon dioxide to manufacture carbohydrates such as sugars and starches via photosynthesis. By about 300 million years ago, carbon dioxide had virtually disappeared from Earth's atmosphere, leaving the nitrogen in the atmosphere as the predominant gas, as it remains today.

Oxygen, essential to animal life on Earth, is a byproduct of photosynthesis that plants then give off. This element is highly reactive, meaning that it combines with many elements and molecules. For hundreds of millions of years, the oxygen molecules released by plants in the young oceans and then by those on the land quickly combined with minerals on the Earth's surface. For example, surface iron was transformed into iron oxide, or rust, by this gas. Once all the places where it could bond on the Earth's surface were filled, oxygen molecules released by plant life on Earth remained in the atmosphere. It is this oxygen, replenished by plant life today, that we breathe.

If all the steps in this transformation process from a carbon dioxide to a nitrogen-oxygen atmosphere occur on Mynoa, then that world will have the potential to support land-based animal life. But did they all occur? The one step that is significantly different there from what happened on Earth, and which therefore requires serious attention, is the formation and persistence of life in the young oceans of Mynoa.

FORMATION OF LIFE ON MYNOA

Once oceans were established, it was essential to add into them the broth of atoms and molecules, called the primordial soup, from which life formed. It was then necessary to keep the primordial soup stirred up so that it did not settle to the bottoms of the oceans and form rock. The process of adding to and mixing the primordial soup on the young Earth was driven by miles-high ocean tides generated by the Moon. These tides rushed inward and scoured the land every few hours. As the tides receded, they brought water brimming with minerals into the oceans.*

The crucial activities of keeping the primordial soup stirred up and preventing it from settling en masse onto the ocean bottom were harder to do on Mynoa. Because that world was in synchronous rotation when the oceans formed on it, they never went through a period of miles-high water tides every few hours, as occurred on the Earth. Recall that from the beginning of its oceans, Mynoa had a tide typically only a few feet high: only one-third as high as we have now. That tide was virtually useless in keeping the oceans stirred up and primed for developing life.

There were at least three other mechanisms available for creating the primordial soup on Mynoa, none as effective as the lunar tides were here on Earth.[†] First was runoff from rivers on Mynoa. If you have ever seen a muddy river meandering down to the ocean, you have an idea of what was taking place in these situations. Rivers carry minerals from the land (the mud) downstream into the oceans, adding to the primordial soup.

Second was debris injected into the water from volcanic sources. At least 5,000 active submarine volcanoes are known here on Earth, along with tens of thousands of dormant ones. In their formative periods both Earth and Mynoa had many more of them. When underwater volcanoes are active, molten rock and ash from underground are injected directly into the oceans. Likewise, volcanoes on land that erupt violently (called stratovolcanoes) cause debris to rain into the oceans. In both cases, some of this material from inside a world dissolves in the water, adding to its primordial soup.

Third were the continued impacts of space debris—asteroids, meteoroids, and comets—into Mynoa's oceans. Asteroids and meteoroids are rock and metal bodies, whereas comets, as we noted earlier, are combinations of ice and rock. Asteroids are larger than meteoroids, which range from dust-sized up to a few yards across. Space buffs refer to any of this material passing through our atmosphere as meteors and any of it that lands intact as meteorites. A meteorite larger than a few yards across will create a crater. For

---

* For further discussion of this, see the book *What If the Moon Didn't Exist?: Voyages to Worlds That Might Have Been*, by the author.
[†] All three of these occurred here on Earth, as well.

example, one the size of a school bus striking the Earth or Mynoa with a speed of more than 25,000 mph will create a crater half a mile across. Much of the smaller debris, vaporized in the atmosphere, falls into the oceans inside raindrops. The larger pieces passing through the oceans and slamming intact into the ocean bottoms create craters. In both cases, debris is dissolved in the water, enhancing the primordial soup. Astronomers have also discovered that comets carry some basic organic compounds which can be deposited into oceans to help enrich their organic content.

These three activities were not nearly as effective in priming the primordial soup on Mynoa as occurred on Earth. The difference is that the river runoff and volcanic activities are both relatively local phenomena, which were not dispersed nearly as effectively by the minuscule tides on Mynoa as was such debris in Earth's young oceans.

Earth's primordial soup was continuously refreshed and global, thanks to the lunar tides. The primordial soup on Mynoa was thin and sporadic, except in concentrated regions along some coastlines and around submarine volcanoes. Granted, ocean currents spread some of this material, but at the same time it flowed horizontally, much of it settled to the ocean bottoms, where it eventually turned into stone.

Energy Sources to Spark Life

Most atoms and molecules don't interact with each other without external energy being applied to them. For life to form in the first place from a broth of these particles, energy has to come from somewhere. On Earth, that energy came from a variety of sources, including lightning, heat from volcanoes and impacts, and probably the most important one, ultraviolet (UV) radiation from the Sun. It is worth noting that near the surfaces of oceans, water allows energy from all these sources to reach the atoms and molecules dissolved in it. Happily, Mynoa had these energy resources, too.

It is useful to understand why ultraviolet radiation is the prime mover in enabling life to begin. Ultraviolet is only one of a wide range of electromagnetic radiations emitted by stars. In order of increasing energy, they are radio waves, microwaves, infrared radiation (which we feel as heat), visible light, ultraviolet radiation (responsible for tans and some cancers), x-rays, and gamma rays.

Atmospheres block virtually all the x-rays and gamma rays from space, which is a good thing because they both pack enough energy to rip apart any biological molecules they encounter. Most radio waves, microwaves, infrared, and visible light pass through atmospheres, but mostly they lack the energy necessary to facilitate the combination of many types of atoms and molecules. A considerable amount of ultraviolet radiation from the Sun or Zon, which has the amount of energy suitable for these processes, passed through the atmospheres of the young Earth and young Mynoa. Both atmospheres lacked the compound ozone (three oxygen atoms bound together), that prevents much of the Sun's ultraviolet radiation from getting to the Earth's surface today. Therefore, ultraviolet provided the most reliable source of energy for particle interactions in the primordial soup.

Based on the arguments above, I posit that all the pieces of geology, chemistry, atomic physics, meteorology, and oceanography necessary for life to form and evolve there eventually came together on Mynoa. However, given the relative difficulty in maintaining the primordial soup, it is likely that it took an extra billion* years for advanced life to evolve into existence there than it did here on Earth.

As with so many other things about Mynoa, its near side and far side experience distinct cycles of ultraviolet radiation. On the far side, whenever the Zon is up, UV from it comes to young Mynoa's surface. That radiation is especially intense when the Zon is high in the sky, as discussed earlier in regard to the seasons. In contrast, there is virtually no ultraviolet bathing the near side of Mynoa during the time when the Zon is being eclipsed by Tyran.

With less energy available to stimulate the formation and evolution of life on the near side of Mynoa than on the far side, I anticipate that life began in oceans on Mynoa's far side. Ocean currents then spread that life, as occurred on Earth, carrying some of it to the near side. One interesting aspect about that spread of life on Mynoa is the impact that the different cycles of light and dark on the near and far sides have on the evolutionary tracks that life takes. This issue is best explored in the context of the biological clocks that this life develops.

---

* This number is intended to indicate the scale of the delay that could occur, rather than to present a definitive interval. Also note that this explains why I considered the location of Mynoa five and a half billion years after it formed as the time when sentient life developed on it, rather than the four and a half billion years that it took on Earth.

Biological Clocks on Mynoa

Life on Earth evolved internal clocks to assist it in responding quickly and efficiently to the activities that occur at different times during the perpetual cycle of day and night in which we are immersed. Your biological clock, which has a roughly twenty-four-hour cycle, is located deep in your brain, in a part of its hypothalamus called the suprachiasmatic nucleus. That clock is responsible for regulating your waking and sleeping cycle, hunger, sex drive, metabolism, and body temperature, among other things. Under normal circumstances, your biological clock is reset each day, typically by the light at sunrise, although changes in pressure and temperature can do the trick. If you have ever suffered from jetlag, you know how hard it is to reset your biological clock after it gets out of synch.* I assume that life on Mynoa also needs to develop biological clocks, but the two different cycles of day and night there will profoundly complicate the process.

As we discussed earlier, the day–night cycle on the far side is the same as the one with which we are familiar (albeit some four times longer): namely, hours of light from sunrise to sunset, followed by hours of darkness through the night. On the near side, we have seen that the Zon is blocked by Tyran during some of the "normal" daylight hours, so the cycle of light and dark is this: light after Zonrise followed by darkness during the Zolar eclipse, followed by light again, followed by Zonset, which leads to a period of changing brightness as Tyran goes through very bright gibbous and full phases, followed by Zonrise. Furthermore, the near side never becomes as truly dark as the far side. All this is summarized in Figure 2.3.

Consider what happens when early biological clock–endowed Mynoan life that formed on the far side wanders onto the near side.† The change in light throughout a day is the major factor that determines the existence and behavior of biological clocks, so creatures from the far side venturing onto the near side would find the midday darkness and midnight light to be serious distractions. Just as do humans who change time zones or who go to Antarctica, these Mynoans will not function well in the screwy cycle of light

---

* People who live in regions of prolonged daylight and darkness are prone to problems related to their biological clocks not being resynched every twenty-four hours, including depression, sleep problems, and seasonal affective disorder, as well as secondary problems such as alcoholism and drug abuse, among others.

† These could be anything from bacteria to plants to fish to mammals.

and dark they encounter. Indeed, most of them will die prematurely because they will be sleeping when they should be having sex, trying to mate when they should be hunting, or doing any number of other things at the wrong times. The good news is that evolution, being what it is, is very likely to cause mutations that enable some creatures swept onto the near side to evolve biological clocks that can accommodate the bizarre near-side cycles of light and dark.

One consequence of the evolution of creatures with fundamentally different biological clocks is that Mynoa will eventually harbor two incompatible families of life-forms. Before sentient creatures (here used to mean beings that are aware of themselves, as we are) evolved among them, this would not be a big problem; the far-side creatures that wandered onto the near side and had to compete with the native near-side creatures would quickly die out because the far-siders would be unable to function as effectively as would the natives, and vice versa. However, when sentient life developed and started exploring the opposite sides of their world, serious problems would develop.

Let's assume that sentient life evolves on both sides of Mynoa, possibly from completely different species and each with its own specialized biological clocks. Early explorers from one side going to the other side quickly understand that they aren't functioning well, and perhaps they even intuit why that is happening. If we humans are any examples, they will eventually develop either technology or medicines to enable them to cope on the opposite side of their world. Then, with two sets of sentient creatures vying for the same real estate, conflict would begin.

TEMPERATURE VARIATIONS

Far Side

By the time sentient creatures live on Mynoa, when the day is about 100 hours long, the daily weather cycle will be very different than anything we experience. The basic factor is that more hours of heating and more hours of cooling occur each day on Mynoa than on Earth. As a result of the cycle of heating and cooling that we experience daily on Earth, the average daily temperature change here is about 20°F. With four times as many hours for heating and cooling, and allowing for the fact that the cooling rate isn't constant throughout the night, Mynoa's daily far-side temperature range would be roughly 40°F.

Perhaps the most significant criterion for the kinds of life-forms capable of flourishing in any area is whether water freezes and thaws there. Consider a typical scenario on Mynoa's far side at a latitude where the noontime temperature is a balmy 65°F. That night, the temperature would plunge below the freezing point of water, remaining there for the better part of fifty hours. During that time, shallow water bodies would radiate away the heat they received from the Zon during the preceding daylight hours and begin freezing. Precipitation during the night would come down as snow. After Zonrise, the ice and snow would melt, with much water evaporating and creating morning mists.

This wide range of temperatures and the changes caused by it would require that life be able to respond to a much wider range of daily temperature than does any life on Earth. Clearly, cold-blooded animals and many types of plants that live on Earth would be unable to withstand the temperature changes that occur at that latitude on Mynoa.

Now let's look at a place where the temperature never falls below freezing. Allowing for a 40°F daily temperature cycle on the far side, daily highs would have to be over 72°F. If cold-blooded creatures exist at all on Mynoa, this is the place for them. Although they would not freeze, such creatures would still have to endure the wide range of temperatures. For example, alligators would be vigorous at 75°F, but they would be immobile at 35°F, and therefore at night they would be easy prey for warm-blooded creatures. As you can see, life evolving on Mynoa will face some challenges that life here did not encounter.

Near Side

Tyran will be a weak heat source for the near side. The planet, heated by the Zon, as well as from radioactive elements inside it, will have an upper cloud temperature of around 80°F. Because of the separation between Tyran and Mynoa, however, the larger body will only provide a low level of heat to Mynoa's near side. Indeed, the near side of Mynoa will be cooler overall than its far side because of the eclipses on the near side: these daily eclipses lead to cooler daylight hours than occur on the far side. During the eclipses, heat from the near side that is radiated into space is not replaced by an equally great amount of heat from Tyran. The resulting cooling of Mynoa also occurs on Earth during solar eclipses.

Mynoa's Magnetic Field

Mynoa's slower rotation has other consequences in addition to wider ranges of temperatures than we experience on Earth. One of these differences is that Mynoa will have a weaker magnetic field than does our planet.

A planet's magnetic field is generated as a result of charged particles moving inside it. In the cores of Earthlike planets there is molten metal that carries charged particles, electrons, upward and downward as the metal heats and cools. I will have more to say about this cycle of motion, called convection, later. Combining this radial (upward and downward) motion with the planet's rotation leads to the kind of magnetic field that Earth has, with north and south magnetic poles, like a bar magnet.

The slower a planet rotates, the weaker the magnetic fields it generates. Mynoa's 102-hour synchronous rotation around Tyran causes the smaller body to have a magnetic field only about a quarter as strong as ours. The consequences are colorful. To understand why, let's briefly review the effects of the magnetic field surrounding our planet. The Sun is continually giving off a spray of charged particles, mostly protons and electrons, collectively called the solar wind. Traveling toward us, some of these particles encounter the Earth's magnetic field. Whenever a charged particle crosses a magnetic field, that field deflects the particle. Charged particles heading in Earth's direction from the Sun are forced by our magnetic field to change direction and thereby miss hitting the Earth's atmosphere. (This is a good thing because the solar wind particles have enough energy to kick particles in the Earth's atmosphere into space. Occurring for billions of years, these impacts would thin the atmosphere.)

Many of these deflected particles become trapped by the magnetic fields in regions surrounding the Earth called the Van Allen belts. These are two flattened donut-shaped regions surrounding planets like Earth and Mynoa in which electrically charged particles from the Sun and elsewhere are bottled up. They are named after James Alfred Van Allen, whose instruments on the *Explorer 1* rocket in 1958 first detected the trapped particles in the belts surrounding our planet. Mynoa's weaker Van Allen belts would allow more solar wind particles to pass right through them and strike the planet's atmosphere. Although this would thin the air there more than occurs on Earth, life on Mynoa would still be able to evolve, all other things being equal.

As well as knocking some gas into space, particles from the Sun cause other gases in the air to glow. Particles leak out of Earth's Van Allen belts from time to time and head earthward. They slam into the gases of the atmosphere, causing some of them to glow, thereby creating auroras. There are also periods of time when the Sun emits bursts of particles that are much more energetic than the normal solar wind. These flares and coronal mass ejections send out particles with so much energy that they tear right through the Van Allen belts and strike the Earth's atmosphere, creating especially impressive auroras.

We now have the background to understand why Mynoa's nights will be more colorful than ours. Mynoa's weaker magnetic field will be unable to trap as many of the particles passing through it as do our Van Allen belts. Therefore, energetic particles from the Zon will continuously strike Mynoa's atmosphere, causing it to glow. The night sky on the far side of Mynoa will often be filled with a psychedelic tapestry of multicolored, ever-changing pastel light.

My own experience with auroras is that they can be mesmerizing. In 1969, as I was driving back to college in central New York State, a curtain descended in the sky in front of my car. I stopped and got out. Silky, gray, and pleated, this aurora hung absolutely fixed for at least half an hour. On October 28 through October 30, 2003, the Sun emitted a series of flares and coronal mass ejections that caused auroras for several days. The ones I saw, on Halloween of that year, filled the sky with bursts of green, red, and white light. There was so much activity that Chicken Little came to mind.

Tyran's Comets

Auroras last for hours or less. Comets are longer-lived, transient phenomena that will be visible from Mynoa relatively frequently compared to what we see from Earth. Throughout its history, comets are also more likely to strike Mynoa than they are to strike the Earth. Comets, we noted earlier, are dirty icebergs in space typically measuring a few miles in diameter. Observations reveal that two reservoirs of comets exist around the Sun and also, we are beginning to observe, around other stars. The first assembly of comets in our neighborhood is in a bagel-shaped volume of space centered on the Sun and extending outward from about the distance of our Neptune. The bagel, formally called the Kuiper belt, after American astronomer Gerard Peter Kuiper (1905–1973), is oriented so that the cream cheese in the middle is the plane of the ecliptic.

The other source of comet bodies is located in a spherical volume of space also centered on the Sun and extending perhaps a quarter of the way out to the nearest star. Called the Oort comet cloud, it was named after Dutch astronomer Jan Hendrik Oort, who predicted its existence in 1950.

Frozen comet nuclei swarm by the billions around stars like the Sun. As such an object approaches a star, heat and the outflow of particles from the star's atmosphere cause ices and dust on the comet nucleus to vaporize, creating its distinctive atmosphere and tails. Receding from the star, these features diminish and often vanish. Most comets that we see maintain their tails for a few years before they move so far from the Sun that there isn't enough sunlight hitting them to free the material that the tails comprise.

The twist in the story of comets and Mynoa is that some comet nuclei will actually be captured in orbit around Tyran, as they are by the giant planets in our solar system. Perhaps the most famous such comet here was Shoemaker-Levy 9, which orbited Jupiter as a single object until 1992. In that year it passed so close to the planet that the tidal force from Jupiter pulled the comet apart, creating at least twenty-three fragments ranging in size up to a mile in diameter. Remaining in orbit for two more years, this debris slammed into Jupiter in 1994, creating ripples and dark regions in Jupiter's gaseous/liquid outer layer. The temperature at the impact sites rose from a normal −225°F up to 40,000°F. Many of these impact features were larger than the Earth. Put another way, each of the impacts had much, much more energy than all the humanmade explosives ever created.

Consider a comet that was orbiting for billions of years in the Zon's Kuiper belt. One day it is struck a glancing blow by another comet that deflects it Zonward and toward Tyran. By the time the comet is captured by the gravitational attraction of that planet, it will have been heated so much by the Zon that it will have a tail. Once in orbit around Tyran, that tail will persist for hundreds or even thousands of years. This will occur because the comet remains so close to the Zon that the dust particles and volatile (easily vaporized) ices it contains are continuously vaporized and blown away.* As a result, any captured comets will be seen from Mynoa day in and day out for centuries.

The shape of the comet's orbit around Tyran is likely to be quite elliptical (elongated) because the comet came from far outside Tyran's orbit. In the usual scenario, the comet will pass only a few thousand miles from Tyran during part of its orbit, and recede a million miles or more when it is farthest away. Mynoa, in its nearly circular orbit, will rarely be near the comet.

Suppose comet Comins (comets are named after the people who discover them) came into orbit around Tyran when prescientific sentient beings lived on Mynoa. Because the comet's tails always point away from the Zon, and because Mynoa and comet Comins have different orbits, the people watching the comet night by night from the far side will see it change shape and move smoothly through the background stars. Sometimes they will see the comet broadside as a long arc in the sky; sometimes they will see it nearly head-on as a giant hazy sphere.

In contrast, people on Mynoa's near side will see it much less often, in part because there are few hours of sufficient darkness (during the daily Zolar eclipse by Tyran) and in part because the comet is usually far from Tyran, where generally it cannot be seen at all from the near side. Seeing it less often and not in a smooth sequence of orientations, it will be harder for people on the near side to detect the cycle of change for the comet's tail than it will for their "cousins" on the far side.

One can imagine a variety of mythologies that would be associated with comet Comins, but I am sure that foremost in all Mynoan minds would be the question: will it ever endanger us? Prescientific Mynoans, watching the comet hang fire in their sky, worrying about its motion and how it changes over the years, not knowing whether it would ever threaten them, are likely to want to understand everything they can about it. This could motivate them to develop science and technology centuries or millennia before we

---

* After all the volatiles have been freed from it, comet nuclei lose their tails and spherical atmospheres, called comas. Thereafter, they shed little matter.

did. The science of astronomy on Earth began for much less compelling reasons, such as to answer the questions of why some objects in the sky appeared forever fixed relative to each other (the stars), why some celestial bodies (i.e., the planets) wander among the fixed stars, and when the seasons will occur.

I suspect that it would be easier for the people on the far side of Mynoa to work out the equations describing the comet's motion, namely those equations developed by Kepler and Newton, than it would for people on the near side. The difference lies in the patterns of motion of comet Comins that would be clear to the far-siders long before it would be understood by the near-siders.

Supposing the far-siders did work out the equation of gravitational force necessary to describe the motion of comet Comins, would they discover that it is a threat? The answer is a definite maybe. Because of mathematical chaos, it is impossible to calculate the orbits of small objects such as the comet far into the future under the combined influences of Tyran, Mynoa, the Zon, and other bodies orbiting nearby. The same applies to small objects orbiting our Sun and passing near Earth. Calculations may show that on the next encounter that object will hit us, but in the intervening years, tiny gravitational tugs of other objects in the solar system could well divert the intruder. The discovery by far-side Mynoans that a collision with comet Comins is likely could well spur development of the technology necessary to prevent it from happening.

A FEW CHARACTERISTICS OF LIFE ON MYNOA

It is unlikely in the extreme that sentient beings on Mynoa or on any other world discussed in this book resemble us, as noted in the Introduction. There are just too many steps in the evolutionary process, too many environmental differences between those worlds and the Earth, and too many places where chance and chaos come into play for that to happen.

Life that can exist in a given environment evolves as a result of myriad random, genetic changes. Most such changes lead to the death of the mutant. Those few changes that make creatures more competitive than their ancestors can lead to competition between the two versions of the species. As a result, the older version of the life-form is sometimes replaced and the species evolves. But we are talking about living creatures and this ideal scenario of improvement and replacement is by no means a sure thing. The improved mutant might catch a cold and die or be accidentally stepped upon or intentionally killed by others in its group or drowned or face any number of other deadly fates before it can reproduce, leaving a status quo where improvement would have otherwise occurred. To put it most starkly, if two identical Earths were formed in an identical astronomical environment, the chances that the peoples on the two worlds would look or behave similarly to each other are minuscule. Nevertheless, there are some things that I (as a scientist, rather than a science fiction writer) can say about life on worlds such as Mynoa.

Carbon

Carbon is the backbone of all life on Mynoa, just as it is on Earth. Backbone, in this context, means a type of atom that can make the complex organic molecules essential for life to exist. The backbone element must be able to bond covalently with three or more other atoms. Covalent bonds are ones in which atoms share electrons. The other type of chemical bond in nature, called ionic, occurs when one atom steals an electron from another and the two atoms stay weakly bonded to each other as a result. Only covalent bonds have the ability to make the suitably strong connections needed for organic molecules.

Why are three bonds essential in creating these molecules? Three is the minimum number of connections that can allow atoms to create molecules of unlimited size and complexity. Let's see what happens to atoms that can bond with only one or two other atoms. Once an atom that can only bond to one other atom has made that bond, then it cannot bond with any other atoms (Figure 2.4a): no life in an H_2 hydrogen molecule.

An atom that can bond with two other atoms can make strings of atoms (Figure 2.4b), but it cannot make more complicated combinations, which we know are essential to the existence of nearly all the organic atoms in our bodies (Figure 2.4c). In fact, carbon can bond with four other atoms

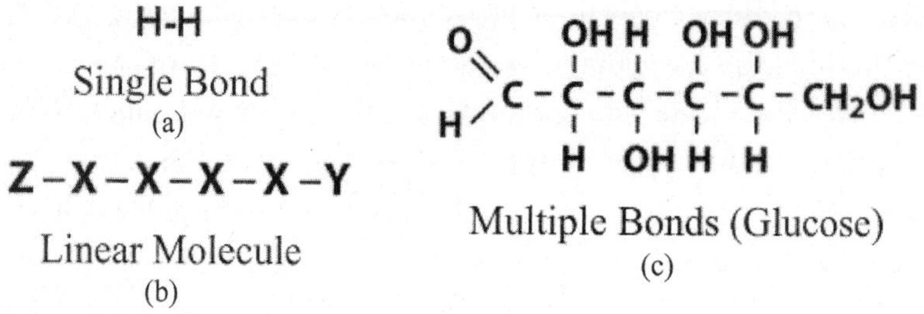

H-H

Single Bond

(a)

Z – X – X – X – X – Y

Linear Molecule

(b)

Multiple Bonds (Glucose)

(c)

Figure 2.4: (a) Single bonds allow only pairs of atoms to connect; (b) a linear molecule; (c) an organic molecule. Organic molecules connect each carbon atom with up to four other atoms or molecules. CREDIT: W.H. FREEMAN & CO.

or molecules, which makes it even easier for complex molecules to form around it.

What about other atoms serving as the backbone for life? There are only four other elements that can make three or more covalent bonds: boron, nitrogen, silicon, and phosphorus. However, experiments and theory show that the bonds these four atoms make are either too inflexible or too flexible to be useful in forming life. Consider silicon, for example. When it bonds with two oxygen atoms, a common activity in and on the young Earth, it forms silicon dioxide, which you know in the form of sand or quartz. In other words, it forms very rigid compounds: rock. Conversely, when silicon combines with one oxygen atom and other molecules in long repeating chains, it forms silicone, which is very soft and pliable. In general, these other four candidate backbones all make compounds that are typically too rigid or too yielding to serve as the basis for life. Carbon alone allows for suitably strong, but flexible and incredibly varied, molecules. Making matters even better, carbon is very common on terrestrial planets such as Earth and Mynoa.

Oxygen

We saw earlier how Mynoa's and Earth's atmospheres became enriched with oxygen. Let's now consider this gas's role in life and whether it is likely that Mynoans use it as we do, or whether there are alternatives. Complex life on Earth requires energy to be stored in each cell in order for the cells to function. The primary and most efficient energy supply source we have is a molecule called adenosine triphosphate or ATP. When the energy stored in this molecule is used, it is converted to the related molecule adenosine diphosphate (ADP), which is then converted back to ATP with the aid of other molecules such as sugar and oxygen. We breathe in the necessary oxygen and have it delivered to our cells by the hemoglobin in our blood. Oxygen's crucial role is to take up electrons in the conversion process of ADP to ATP. Furthermore, carbon dioxide is one of the waste products in this process.

ATP can be made without oxygen, but the processes to do that are very inefficient compared to those using oxygen. Therefore, I think it is likely that the energy-supplying processes for life will use oxygen on Mynoa, as they do on Earth. There are other energy-rich molecules that can be used to power cells, but once again, re-energizing them is likely to involve oxygen. Therefore, even if animal life on Mynoa doesn't evolve the biological machinery to manufacture and use ATP, it is extremely likely that oxygen will still mediate generation of the energy-rich cells used to power life. I therefore anticipate that oxygen will be an essential molecule for sustaining animal life on Mynoa.

If we accept this conclusion, that doesn't mean the Mynoans will have noses and mouths that take oxygen into lungs and expel carbon dioxide as we do. Specialized organs such as lungs are particularly efficient devices to exchange gases, but even here on Earth, there is a variety of ways to get these gases where they are needed, such as gills and openings in the skin called spiracles, used to transport gases in and out in the tracheal system in insects. This latter plumbing is not nearly as efficient as are lungs and a circulatory system to carry the oxygen in and carbon dioxide out, such as the ones we have. Likewise, iron-rich hemoglobin is not the only oxygen-carrying molecule that life could use. Therefore, Mynoans' blood need not be red.

Photosynthesis

Although animal life can generate and maintain its cells using internal energy sources such as ATP molecules, plant life needs external energy such as sunlight to keep it alive. On Earth, that energy conversion process in plants is mediated by chlorophyll molecules. Absorbing a lot of red and blue light, they use that energy to convert water, atmospheric carbon dioxide, and other molecules into the compounds needed to build and sustain plants. Most plants look green because their chlorophyll-rich leaves don't absorb much green light, which is therefore scattered away from the plants. If more efficient molecules exist in the plants on Mynoa (or if those plants use chlorophyll plus other molecules), then the plants there may well have leaves of different colors. If they contain molecules that absorb more of the visible colors, leaves on Mynoa could well look darker or even black. Instead of indicating dead plants, such black leaves would be a sign of their vigorous absorption of Zolar energy.

Vision and Hearing

Just as on Earth, life on Mynoa will need to sense its environment in order to survive. Different senses provide information about what is happening at different distances from you. For example, your senses of touch and taste provide information about things with which you are in contact, smell and sound tell you about nearby events, and sight gives information about things from inches to trillions of miles away.

In order to refine sensory information, namely to tell distances to events of interest, life has evolved pairs of sensors for both sight and sound. A single eye can tell you the direction of a fire, say, but it can't tell you the distance to it. For accurately measuring distances, animals need two eyes so that they and we can triangulate on objects we see. In other words, two eyes looking at the same thing see it at different angles (Figure 2.5). The brain uses the different angles at which the eyes are looking to tell it how far away objects are. Therefore, we can reasonably expect that creatures on Mynoa with eyes (or their equivalent) will have at least two of them. We could imagine that some animals there might need to see all around themselves all the time, meaning that they might have three or four eyes, but at least two. A similar argument can be made for ears: two are necessary for quickly and accurately determining the direction, if not the distance, to a sound.

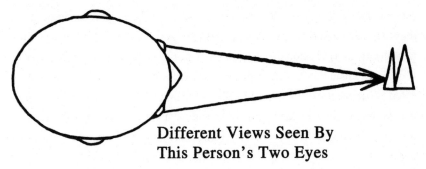

Different Views Seen By This Person's Two Eyes

Figure 2.5: Parallax: seeing the same object from different angles enables the brain to determine the distance to the object.

ASTRONOMICAL MYTHOLOGY ON MYNOA

The Night Sky

As we noted earlier, people on Mynoa's far side will see a sky as ebon as the darkest nights seen on Earth, a sky ablaze with thousands of stars. Even better, because their sky is never brightened by a moon or by Tyran, they will see these spectacular displays of stars every clear night. We, in contrast, rarely see many stars or the silky glow of the Milky Way because our Moon is so bright through most of its phases; moonlight obscures the light from dim objects in space. The best time on Earth to see the kind of night sky Mynoa's far-siders see every night is during the new Moon phase (and far from city lights).

In contrast, people on the near side of Mynoa will only have a few hours of significant darkness each day and even then their sky will never be the deep black seen from the far side. The stars will be brightest during the daily eclipse of the Zon; after Zonset, the light scattered off Tyran will make the sky much brighter than does moonlight on Earth.

Although the glimpses of stars will tease the near-siders, people on the far side will be able to kick back and watch the stars and their patterns at their leisure. The far-siders will see that most of the stars remain fixed relative to the other stars throughout their lifetimes and, in all likelihood, throughout recorded history. They will see a few wanderers in the sky that are in different places among the background stars each night. They will probably call them the equivalent to planets (Greek for "wanderer"). Perhaps most important, they will see patterns of stars in the sky, which we call asterisms or, more commonly, constellations.

Constellations exist because stars have different brightnesses and are located at different distances and in different directions from us. Even though these stars are distributed by natural forces, we see patterns of them as symbolizing objects from our societies. This pattern recognition behavior occurs because the human mind evolved so that we can quickly determine whether things we see are friends, foes, or food. For example, the sooner we perceive a person wielding a weapon as a potential threat, the sooner we can respond appropriately.

So powerful is the pattern recognition hardware in our brains that we see patterns even when they don't exist. That is why virtually every culture has identified patterns of bright stars as objects that are important to them. For example, the asterism we call the Big Dipper in the United States is called the Plough in the United Kingdom, the Butcher's Cleaver in northern England, the Casserole in France, the Seven Sages in India, and the Northern Dipper in eastern Asia, among other names. J.R.R. Tolkien incorporated the Big Dipper as a Sickle into his Middle Earth mythology. The same analysis can be done with the names of many of the other prominent asterisms, despite the fact that the patterns of stars have nothing to do with the objects for which we have named them.

Mynoans will see patterns among the stars; however, they will not see the same patterns that are visible from Earth. The reason for this difference is that at different places throughout the Milky Way Galaxy different stars are visible. Unless Mynoa is located exactly where we are in the Milky Way Galaxy and it is here today, they will see different constellations. Even here on Earth, over times as (astronomically) short as a few tens of thousands of years, the shapes of the asterisms change. Figure 2.6 shows what the Big Dipper looked like 50,000 years ago, how it looks today, and how it will look 100,000 years from now. Therefore, cosmic mythologies on Earth created by constellations will change with time. Because they see different patterns, the names and mythologies that the Mynoans develop about the bright stars are assuredly different from the stories we have devised for the patterns we see today.

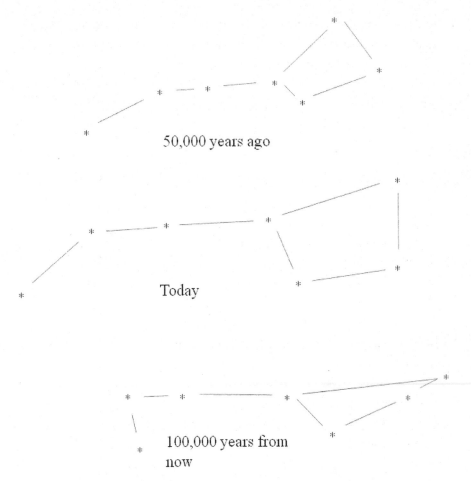

50,000 years ago

Today

100,000 years from
now

Figure 2.6: Changing constellations. Ursa Major (The Big Dipper) as it appeared 50,000 years ago, today, and 100,000 years from now.

Building upon the Sky

Astronomy, and before it astrology, played important roles in the history of humankind. The cycle of day and night is measured by the location of the Sun. The time of year can be determined by what stars are up at night. The phases of the Moon, lunar and solar eclipses, comets, supernovae (exploding stars), shooting stars, the nightly motion of the planets among the fixed stars, auroras, and even giant groups of sunspots visible to the naked eye,* have all

---

* Never look at the Sun without approved eye protection. Doing so causes blindness.

intrigued people since time immemorial. In prescientific times, astrologers associated such events and their locations in the constellations with human activity. Many of these events will add to the lore of the people on Mynoa's far side, but I think it is very likely that near-siders will have even more interest in the events in the sky and their association with life on their planet. That interest will stem from the presence of Tyran.

The Mystique of Tyran

Imagine living on the near side of Mynoa, seeing huge Tyran fixed in the sky. That planet has parallel stripes with different hues of azure blue, sweeping majestically around the planet's poles. Between the belts of color are swirling patterns of roiling gases. Auroras appear, sometimes flickering in and out of existence, sometimes sitting above the planet like crowns or curtains. Wispy clouds of white and light brown hues develop, often spreading into thin lines, like clumps of unmixed batter in a blender. Every day Tyran goes through a cycle of phases. Why does all that happen?

Also, on the near side, during the brief hours of darkness, stars will be seen falling out of the sky (shooting stars, i.e., meteors). These will be so rare that most people will never see one. Lightning has not been confirmed on Neptune, but it is likely that Tyran, being much closer to the Zon than Neptune is to the Sun, will have it. Living on Mynoa, you see the flashes in Tyran's clouds, but you never hear the accompanying thunder, as you do during thunderstorms on your world. Why not? Comets occasionally fly into Tyran, creating monster ripples that spread over large areas of that world, but again, you hear no sound from these events. And why in the world doesn't that monster body come crashing down onto Mynoa? After all, it's just hanging there. Or perhaps it will fly away!

Then, as is "human" nature, you begin to compare Tyran to your home world of Mynoa. What makes it look so different with its veil of ever-changing clouds through which no Mynoan has ever seen land or oceans? Why aren't there at least mountains peeking through Tyran's clouds? Where do the life-forms on it live? After all, in some ways it is an active world like your own so it must have life (not really, but you as a Mynoan don't know that). What are Tyranese like? Do they have the ability to see through the clouds and know that you are out there watching them? Will they ever come and visit? Will they be friends or foes?

Whereas systematic motion of objects such as comets will intrigue far-siders, the innumerable questions that Tyran generates in the minds of the near-side Mynoans, especially those questions related to their safety, might well motivate near-siders to develop what we would consider to be a scientific understanding of gravity, orbits, planets, motion, and numerous other physical things thousands of years earlier than did our ancestors. After all, the near-side Mynoans would consider it a necessity, rather than a luxury, to have the equations and physical insights, not to mention the hardware, necessary to deal with perceived threats from Tyran and its life-forms.

I have presented different motivations for the peoples on the near and far sides of Mynoa to learn science and, hence, move into technological eras. The far-siders see systematic behaviors in the night sky, such as the motion of the planets and captured comets. Trying to understand these activities could drive discovery, as they did on Earth. The near-siders see the intriguing Tyran. Which group is likely to advance more rapidly? I believe that the near-siders will develop science and technology faster than the far-siders. The driving force will be fear of the potential effects of Tyran and the Tyranese, although Mynoan historians will likely present the discovery process as driven by curiosity.

Let's briefly consider what will happen when the near-siders and far-siders on Mynoa meet. Although the complete answer is beyond the realm of science there are some factors worth noting. One of these is that the races have different biological clocks, mentioned earlier. Another factor is disease.

Diseases

In 1634, as many as ninety-five percent of the native Americans living along the Connecticut River in New England were killed by smallpox brought there by Dutch traders. The list of similar plagues caused by a people being exposed to a disease for which they had no resistance is legion around the world. Analogously, prior to their meeting and interacting, people on Mynoa's near side are unlikely to have resistance to diseases that the far-siders harbor, and vice versa. It seems likely that epidemics, similar to the ones that occurred on Earth when different races met, will also occur on Mynoa.

Sex

Today some anthropologists and other scientists are exploring the possibility that Neanderthals and humans successfully interbred. If they did, it implies that Neanderthal genetic makeup was exceptionally close to ours. Because the different subspecies of humans evolved from a common genetic source, this is at least plausible. What about on Mynoa? Assuming that the near-siders and far-siders also evolved from common ancestors, will they also be able to interbreed? I won't propose an answer, but keep in mind the genetic differences that we have explored between the two groups: they evolved different biological clocks and they developed different capabilities of withstanding temperature changes each day. There are undoubtedly other biological differences between them caused by different evolutionary tracks.

A Goal for Space Travel and a Matter of Some Gravity

With the advent of the technology for space travel on Mynoa, the first goal would obviously be to reach nearby Tyran. By the time that becomes possible, the near-side Mynoans would also have the technology and scientific background necessary to understand the basic properties of Tyran, such as its mass, diameter, and overall chemical composition, just as we knew those properties of Neptune before the *Pioneer 10* spacecraft visited it in 1983. Nevertheless, or possibly because of this knowledge, Mynoans would launch robot probes into Tyran's atmosphere. These would return information about the chemistry, temperature, pressure, wind speeds, and radiation levels below the clouds.

The spacecraft would also reveal the complete lack of a solid surface on Tyran, the complete lack of life there, and the complete inability of Mynoans to land or live on that giant world. The same is true for us and our Neptune. This raises an intriguing question that we humans didn't have to face: if, unlike our Moon, Tyran cannot be a target for early Mynoan space travel, and because all other objects in the solar system are too far away or too inhospitable for their early explorers to visit, where would they go when they begin leaving their planet?

To add insult to injury, the presence of Tyran will make the exploration of space much harder than it has been for us in another way. In order for a spacecraft to leave the vicinity of Mynoa, it must have enough speed (or equivalently, energy) to overcome the gravitational bonds of its home world. The same applies, of course, for spacecraft we send out from Earth to explore or observe other worlds. The problem is that to leave the vicinity of Mynoa, a spacecraft would also have to overcome the gravitational attraction of Tyran. Considering that the larger body has seventeen times the mass of Mynoa, it will take considerably more powerful rockets for Mynoans to send satellites to other worlds than it does for us. They would have a much harder time leaping into the cosmos than we are having.

3

What If the Moon Orbited Backwards?

ANILLO

The lecture hall filled quickly with the greatest scholars of the day. The older men all sported beards, and the younger ones wore mutton-chops. The semicircular tiers of dark wooden seats above and surrounding the lecture floor filled to capacity, and beyond. Everyone glanced periodically at the three empty chairs on the stage.

As the clock chimed two, the door behind the lectern opened. A blond man in his mid-thirties walked out followed by two children, a boy of ten, skinny, with dark hair that had been slicked down but was already in the process of springing back out of control, and a girl of twelve, pretty, redheaded, and beginning to show signs of woman-hood. The children sat down, while the man walked purposefully to the podium.

"Thank you for coming here today. My name is Leon Finnel. These are my children, Meredith and Laurence. They are both," he paused, "exceptional thinkers, as you may know from their publications in the Proceedings of the Royal Society. We are here today because they have different ideas about the fate of the world and we would like your help deciding who is correct."

The children looked at each other and Laurence stuck his tongue out at his sister. She kicked him, a move barely visible behind the folds of her long dress.

"Laurence will speak first, as his ideas are more straightforward."

Laurence waited while his father put a platform behind the lectern. Then the boy mounted it and looked out at the audience.

"We all know," he began, in a high-pitched, child's voice that caused many members of the audience to laugh. Laurence stopped and looked around, his brows coming together. He waited for silence and began again. "We all know that the Noom is coming toward us."

"Says who?" "What gives you that idea?" and similar shouts drifted down from the viewers.

Laurence looked at his father. "Do I have to talk to them? They're dumb."

"No, they're not," Leon said in a stage whisper. "They just haven't thought about this like you have and Meredith have." Then, turning to the guests, he said, "Sorry. He's just a boy." Laurence stuck his tongue out at his father.

Turning back to the audience, Laurence continued. "I hope you know that the Noom causes most of the tides here on Anillo," he said, looking around for confirmation, which came slowly, hesitantly. He looked back at his father with a "can they really be this dense" look.

"Look," he said, "if you apply Newton's inverse square law of gravity, take its gradient, and work out the motion of our planet and the Noom around their center of mass, you will see that there is a direct correlation with the location of the Noom and the location of the high tides."

"What about the Sun," someone called out.

Laurence brightened. "Yeah, it's involved, too. But it creates less than a third of the tides. We can ignore it for now." He looked around, but no one objected. "Here's the important stuff. The Noom is orbiting in the opposite direction to which Anillo spins. Therefore, the high tide closest to the Noom is pulled around behind it by the planet. This tide is therefore pulling backward on the Noom, which loses energy and angular momentum and so it is spiraling in toward us. It's going to hit us in about two billion years."

Pandemonium filled the room. "Rubbish!" "That's crazy!" "Stupid kid." "Who really does the work they publish?"

Laurence raised his hand and the adults, seeing him standing so, quieted down.

"Let me show you," the boy said, quietly, clearly unsettled by the response to his statement. He took a piece of railroad chalk from under the chalkboard behind him and, moving the platform on which he stood to it, began making drawings on the board, with details such as angles, distances, rotation and orbit rates, as well as equations, which he explained carefully, especially because many people in the audience demanded to see their derivations.

For over an hour he kept the raucous scientists, philosophers, and theologians in their seats, many taking careful notes. Leon noted with satisfaction that the more Laurence talked, the more the scholars began nodding their heads in agreement and approval. When Laurence was done, there was a smattering of applause, intermixed with muttering and spirited conversation between members of the audience.

"My daughter, Meredith, has been working on this separately. She wants to show why Laurence is wrong," Leon said from the podium. He flipped the chalkboard over, stepped aside, slid the platform back in place and waited until Meredith mounted it before taking his seat. In the meantime, several members of the audience lit their cigars. By the time Meredith was ready to speak, the audience was a blur through the thick blue haze of smoke. She looked up, opened her mouth to speak, and began coughing and then wheezing.

"Daddy," Laurence said, urgently. Leon was on his feet, sweeping his daughter into his arms, running for the door.

"Asthma," he called behind him. "The smoke. We'll have to do this some other time." Laurence rose and followed them out.

In the first two chapters we explored two variations on the theme of moons: Earth with two moons and Earth as a moon. Two other lunar scenarios would profoundly affect Earth and life on it: Earth with no moon, and Earth with a moon orbiting in the direction opposite to the direction that the planet spins. I explored Earth without a moon in the book *What If the Moon Didn't Exist?* The present chapter explores the latter scenario. The planet is called Anillo and its one moon, Noom, orbits in the opposite direction to our Moon.

Anillo formed with the same physical and orbital properties possessed by the Earth prior to the impact that created our Moon. The star plays only a minor role in this scenario, so I just call it the Sun. As explored in Chapter 1 and the Appendix, astronomers have suggested four mechanisms for the formation of a moon: co-creation with the planet, fission from the planet, impact off the planet, and capture.

We can easily eliminate the first two as ways in which Noom could have formed orbiting in the direction opposite to which Anillo rotates: terrestrial (Earthlike) planets form as the result of myriad collisions of swirling debris analogous to water in a bathtub swirling down the drain. This debris creates a spinning planet. If a moon orbiting in the opposite direction to that spin begins forming in orbit around the young planet, then some of the debris that would otherwise have hit the planet will collide head-on with the young moon, slow it down, and cause it to fall onto the planet. Therefore, co-creation is not a viable option for Noom.

If Anillo were spinning fast enough to fling debris into orbit, then that debris would have to be moving in the same direction in which the planet is rotating. It's the same effect as a baseball pitcher throwing a ball. As soon as the ball leaves the pitcher's fingers, it must keep going in that direction. It can't suddenly stop, turn around, and head into the outfield before it is hit. Therefore, Noom could not fission off Anillo and orbit that planet in the opposite direction to the planet's spin.

NOOM, ANILLO'S MOON, DID NOT FORM FROM AN IMPACT

Making the case that an impact can't create Noom requires a little more care. We noted in Chapter 1 that most astronomers and geologists believe that the Moon was assembled from debris that splashed into orbit as the result of the impact of a Mars-sized intruder that struck the young Earth. Not only did that impact put rubble into orbit (creating a short-lived ring around the Earth), but it probably also tilted the Earth's rotation axis, thereby creating the seasons, and it changed both the rate that the Earth was spinning and the mass of the young planet.

To understand the effect of that impact on Earth, imagine two cars in a demolition derby. (This is an event in which cars smash into each other until only one is still moving.) If two cars are going in the same direction and the one behind rear-ends the one in front, the one behind gives the one in front a push, causing the front car to go faster. Analogously, the impact on Earth that formed the Moon was in the direction the planet was spinning. The debris from the Earth's surface and from the intruder's surface was mostly ejected in the direction of that spin (analogous to the front car speeding up in the derby), giving the debris the energy needed to boost it into orbit. The debris orbited the Earth in the same direction that the Earth rotates, leading to the formation of the Moon orbiting in the same direction. This effect of pushing impact debris forward is also what happens when you throw a rock in a pond at an angle: much of the water splashes up in the same direction that the rock was heading.

To understand why Noom would not be formed by an impact, consider two cars in the demolition derby that collide head-on. If the two cars are identical and are traveling at the same speed, then upon impact they would crunch together and come to a dead stop. If one had more mass and were moving faster, then it would cause the smaller, slower car to stop and reverse direction, and the more massive one would merely slow down. Likewise, if an impact on Anillo had been opposite to the direction of its rotation, then much less of the debris thrown up by the collision would go into orbit. Rather, most of the debris from such an impact would splash upward at too steep an angle to go into orbit. (This is analogous to cars that hit head-on slowing on impact.) This debris would rise and either leave the vicinity of Anillo altogether or else it would fall back down onto the planet. If the incoming body were massive enough or moving fast enough to splash enough debris to form Noom and put it into orbit opposite the direction of Anillo's rotation, then that intruder would have had to hit the planet with such force that Anillo would have been pulverized in the process. Therefore, I posit that Anillo did not undergo a collision to put a moon in orbit traveling opposite to that planet's spin.

Capturing Noom

As with Lluna in Chapter 1, the only way Noom could come under Anillo's thrall is if the moon were captured. Noom is going to have the same mass as our Moon. The process of capturing such a moon requires a fairly delicate balance of forces, especially when there is no other substantial moon such as Kuu in Chapter 1 that is already in orbit. As we saw, this latter moon would help take energy away from the incoming body, thereby aiding its capture by the planet.

The orbit of a moon going in the opposite direction to the planet's rotation is called a *retrograde orbit*. At least thirty-one moons in our solar system are in such orbits. The one most similar to Noom is Triton, in orbit around Neptune. Triton is three-fourths as large as our Moon and astronomers believe that Triton was captured, just as Noom is about to be.

The way we make the capture physically possible is to have Noom be one of a pair of bodies orbiting each other, as were Lluna and its companion in Chapter 1. Noom and its companion would be flying through the solar system before they encountered Anillo. To be captured, Noom had to shed itself of its partner. This occurred because these two bodies felt different gravitational forces from the planet, a tidal effect that pulled them apart. By separating, the companion body carried away enough energy to allow Noom to go into orbit around Anillo. Where that partner went, we'll never know.

Initially, Noom orbits Anillo at a distance comparable to where our Moon is today. For the sake of simplicity, I assume that Noom's orbit is nearly circular (consistent with the orbit of Triton around Neptune, which was captured and which is orbiting in the opposite direction to Neptune's rotation) and nearly over Anillo's equator, analogous to the Moon's orbit around Earth.

The gravitational disturbances to Anillo caused by Noom and its partner when they approach that planet, as well as when the two bodies separate from each other, and when the partner flies away, will all create significant, but not catastrophic, disturbances on Anillo's oceans. As a result of these changing gravitational tugs, the bodies of water will slosh for about a month, creating tsunamis and Anilloquakes. The time of Noom's capture, $4\frac{1}{2}$ billion years after Anillo formed, is equivalent to the present time on Earth. However, it is unlikely that life will be nearly as advanced on Anillo at that time as it is today on Earth, so that the sloshing will cause less catastrophic damage to life than such an event would have if it had happened when people were on the surface of Anillo. The primary reason for the delay in development of life on Anillo relates to the effectiveness of high tides in filling the oceans with the minerals necessary for life to form and evolve in them, as introduced in Chapter 2.

The chances of life evolving in a thinner primordial soup are lower and hence the events leading to sustainable life there are likely delayed by hundreds of millions of years. I therefore delay the evolution of "people" there by half a billion years. Given that Noom arrived in orbit at the equivalent of the present day on Earth, Noom will orbit Anillo for half a billion years before people will exist to be aware of Noom or to think about its consequences, many of which center around the rate at which the planet rotates.

ANILLO'S INITIAL SPIN

In keeping with Anillo's similarity to the Earth, I want to set its rotation rate to be close to what ours was originally. Although we don't know the Earth's initial rotation rate, we can qualitatively consider how that spin rate changed as a result of the Moon-forming collision. Several factors related to the intruder determine whether its impact caused the Earth to spin faster or slower. These include the intruder's mass, its speed relative to the Earth, the intruder's rotation rate, the Earth's initial rotation rate, and the angle at which the incoming body hit the Earth. The truth is that you can get a wide range of changes in the Earth's rotation rate depending on these parameters, as seen in computer simulations of impacts that have been carried out.

Most combinations of factors cause the Earth's rotation rate after the collision to be higher than its original rate. (If the intruder is rotating—spinning—rapidly in the direction opposite to that of the Earth, then after the impact Earth can in principle end up rotating more slowly.) All these factors contribute to my setting the planets' initial rotation rates corresponding to an eight-hour day and postcollision rotation rates corresponding to five-hour days in the scenarios of this book. Because Anillo never undergoes this collision, its day will never decrease to five hours. At the time of Noom's capture, Anillo's day is still eight hours long.

The other property of the Earth that changed on impact was its mass. This event does not occur for Anillo, so we need to determine what its mass is likely to be based on the Earth's initial mass.

Setting Anillo's Mass

As we discussed earlier, planets build up mass from impacts over billions of years. The impact that created our Moon changed the mass of the Earth. I want to set Anillo's mass to be similar to the mass of the Earth prior to that event. The impact must have been stupendous. The intruder struck the young Earth causing some of the planet's mass and some of its own mass to splash into space. Some of those ejecta were moving so fast that they escaped from the gravitational pull of the Earth–intruder system and left forever. Some of the ejecta went into orbit around the Earth–intruder system and eventually coalesced and became the Moon.

The question becomes: is the total mass of the ejecta greater than or less than the mass of the intruder? If the ejecta mass is greater than the mass of the intruder, then the Earth will have lost some mass as a result of the collision, even when the remaining mass of the intruder settles into it. If the ejecta mass is less than the mass of the intruder, then the Earth will have gained some mass from the event.

Calculations reveal that the Earth gained mass as a result of the impact. The intruder, with the mass of Mars, had nearly nine times as much mass as our Moon. It appears that only twice as much mass as our Moon was splashed off the Earth, so the net increase in the Earth's mass as a result of the impact was about eight percent. Because Anillo never experiences such an impact, that world has the same mass as the Earth did prior to when it was struck by the intruder. Anillo therefore has a mass equal to .92 times (i.e., ninety-two percent) the Earth's mass.

Speaking of impacts, the one that created Earth's Moon also tilted our planet and caused the seasons, as discussed in Chapter 2. Anillo didn't experience a planet-tilting collision, so it seems reasonable that Anillo will not have any seasons. Let's explore why this is incorrect.

Seasons

When Anillo was formed, it was rotating perpendicular to the plane in which it orbited the Sun. That plane of orbit is called the ecliptic. To help visualize this, imagine a table that represents the ecliptic. The planet orbits the Sun along the surface of the table. Take a pencil, representing the north–south pole axis of the planet's rotation, and hold it perpendicular to the table.

This differs from the angle to which the Earth's axis points relative to Earth's ecliptic, as shown in Figure 2.2. Slide the pencil over the table in a nearly circular path, keeping it perpendicular to the table. This simulates the motion of young Anillo's axis around the Sun.

In 1994, astronomers who were modeling the behavior of Earthlike worlds without a substantial moon discovered that over millions of years such a planet inexorably tilts, changing the direction to which its axis points. Sometimes the rotation axis is perpendicular to the ecliptic, as just described for young Anillo, but most of the time it is tilted up to, and even beyond, the tilt of the Earth's axis today. At these latter times, which are most times throughout its existence, a moonless planet like Anillo (before Noom arrives) has seasons. I choose to have Anillo capture Noom when the planet is tilted over at an angle similar to the Earth's 23½-degree tilt.

Once a moon is inserted into the scenario, the tilt of the planet thereafter remains nearly constant. Therefore, Anillo will have the same seasons that the Earth has by the time people evolve onto Anillo.

The Length of the Day on Earth and Anillo

In summary, Anillo's day was initially eight hours long, the planet has about ninety-two percent as much mass as the Earth does today, and before Noom came Anillo had seasons that varied from nonexistent to very large temperature ranges (see, e.g., "What If the Earth Were Tilted Like Uranus" in *What If the Moon Didn't Exist?*). All other things being equal, having less mass implies being physically smaller. Anillo has about ninety-seven percent the diameter of the Earth and the force of gravity you would feel standing on its surface is also about ninety-seven percent as strong as what you feel from Earth. In other words, just by being there, you would weigh three percent less than you do today.

Earth, we have seen, had roughly a five-hour day shortly after the Moon formed. We know that the day has been slowing down ever since, as it is now twenty-four hours long. Due to tides created on it by the Sun, Anillo's day will also slow down even without a moon like ours. How much does Anillo's day lengthen from eight hours over the same 4½ billion years, before it captures its only moon?

Earth's day is lengthening in part because it gives some of its energy and angular momentum to the Moon, causing the Moon to spiral away, as noted in Chapter 1. As we saw there, the Moon is still receding presently at a rate of about an inch and a half per year. Recall that the recession is caused by the tides on Earth pulling the Moon forward in its orbit around us. The energy to do this comes from the Earth's rotation, hence our planet slows down. Put technically, our Moon is gaining angular momentum, which the Earth loses at exactly the same rate.

Angular Momentum

It is worth a quick detour here to explore briefly the concept of angular momentum. If an object is either spinning (rotating) or orbiting another body (revolving), then the object has a property called angular momentum. It is analogous to the linear momentum an object has when moving in a straight line. Linear momentum is simply the product of the object's mass times its velocity in a straight line. The greater the linear momentum, the harder it is to change the object's straight-line motion. The greater its angular momentum, the harder it is to change an object's rotation or revolution. The faster an object rotates, the greater its angular momentum becomes. Similarly, the farther an object is from the pivot point around which it revolves (e.g., the distance from a moon to the center of the planet holding it in orbit), the greater its angular momentum.

A spherical object such as a moon revolving in a circular orbit around a spherical object such as a planet (Figure 3.1) has angular momentum equal to the moon's linear momentum times the distance between the centers of the two objects. The definition becomes more complex as the object changes shape; however, this will do for our purposes.

For a rotating object such as the Earth, the equation for angular momentum is given by replacing the mass with the object's moment of inertia and its straight-line velocity by its angular velocity. The greater an object's moment of inertia, the harder it is to change its rotation rate, just as the greater its mass, the harder it is to change its straight line motion.

The key property of angular momentum is that for a system—say, Earth and the Moon or Anillo and the Noom—isolated from other forces, its

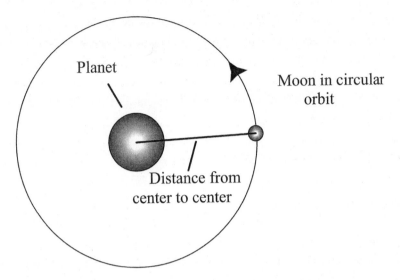

Figure 3.1: The angular momentum of a moon in a circular orbit around a planet is just the moon's linear momentum at any instant times its distance from the center of the planet.

*angular momentum is conserved.** If one body, say, the Moon, gains angular momentum, the other body, Earth, must lose exactly the same amount. As our Moon spirals outward, its angular momentum increases. Therefore, at the same time, the Earth's angular momentum decreases, meaning that the Earth's rotation rate slows.

Another useful fact is that for a fixed angular momentum, the more spread out the mass of a spinning object, the more slowly it rotates. If you were able to bring that object's mass inward without changing the angular momentum, the object would spin faster. Consider, for example, an ice-skater spinning with her arms and a leg extended outward. In order to spin faster, she implicitly uses the principle of conservation of angular momentum by drawing those appendages inward, which brings them closer to her axis of rotation.

The calculation of how much the Moon and Sun each contributed to the tidal slowing of Earth at all times over the past 4.5 billion years is beyond

* They aren't really isolated, because the gravitational force of the Sun is significant, but for our purposes here, we can ignore it.

our present capability because we don't know the details of how the Earth's mass was distributed over the history of the planet. This is because the continents have moved over time and we don't know the details of that motion.

The positions of the continents are important in determining the Earth's rotation rate in part because of the effect they have on the tides. When the continents run north–south for large ranges of latitude, as they do today, the friction between the tides and the land, which causes the Earth's rotation rate to slow down, is much greater than when the continents are consolidated into a large single supercontinent, such as they were as Pangaea 200 million years ago.

Here is a "back of the envelope calculation" about how much Anillo would slow down given that it is identical to the young Earth except that it didn't have a moon for billions of years. When our Moon first formed, it was perhaps ten times closer than it is today. The tides back then were 1,000 times higher than they are now. The height of the tides and the resulting friction between them and the land, which slowed the Earth's rotation rate, has decreased dramatically as the Moon spiraled outward.

After the first two billion years of the Moon's moving outward and the Earth's spinning down, the tides were only a few percent higher than they are now and the Moon was almost out where it is today. During this early time the Sun slowed the Earth's rotation rate further, but much less than the Moon, with the towering tides it generated. For the past two and a half billion years the Moon has slowed the Earth's rotation by just twice as much as the Sun. Combining these data, we find that the Sun has slowed the Earth by about one hour per billion years over the lifetime of our planet.

Anillo started with an 8-hour day. Assuming that it slowed by one hour per billion years, Anillo will have no more than a 12½-hour day 4½ billion years after it formed. With this as background, I set Anillo to be rotating with a 12-hour day when it captures its moon, Noom.

EVOLUTION OF NOOM

We now have a Moonlike moon, Noom, in orbit around an Earthlike world, Anillo. Once in orbit, Noom will start generating tides on Anillo, but with a difference from those our Moon creates on Earth. Imagine viewing Anillo

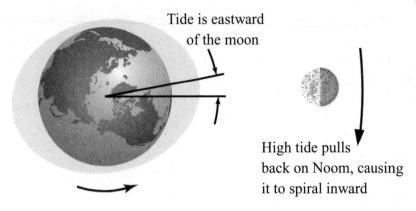

Tide is eastward
of the moon

High tide pulls
back on Noom, causing
it to spiral inward

Figure 3.2: The tides on Earth or Anillo are dragged eastward by the planet's rotation (distances not to scale). CREDIT: W.H. FREEMAN & CO.

from out in space above its North Pole. (This is also the location that astronomers most often use to visualize Earth in relation to its motion through the solar system.) Anillo will appear to rotate counterclockwise once every twelve hours from this vantage point, and Noom will orbit clockwise once every 27⅓ days. Other than going in the opposite direction, the orbit of Noom is momentarily similar to that of our Moon. But its fate couldn't be more different.

Because Anillo rotates very rapidly compared to the motion of Noom in orbit around it, the tides created on the planet by its new moon* will be dragged eastward around the planet, just like the tides on Earth today (Figure 3.2). As with the Earth–Moon system, the high tide closest to Noom pulls on it more than the high tide on the other side of Noom. Recall that this tidal pull from Earth causes our Moon to gain angular momentum and therefore to spiral away from us. Unlike our Moon, which is pulled forward in its orbit by Earth, Noom will feel a force from the nearest high tide that opposes its motion clockwise around Anillo. This force takes energy and angular momentum away from Noom and, as a result, the moon starts spiraling inward, toward the planet.

The fact that Noom is losing energy and angular momentum intuitively suggests that it will slow down as it gets closer and closer to Anillo. In fact, just the opposite happens: it speeds up! If you have ever put a coin in one of

---

* Not to be confused with the moon's phase.

those cone-shaped donation devices in which the coin spirals inward to a hole in the center, you have seen an analogy to what is happening to Noom. By the time the coin reaches the hole, it is spinning around much faster than when it started. Likewise, Noom spirals inward and its orbital motion speeds up.

Noom Watching

As Noom spirals inward, the tides it generates on Anillo will increase, thereby enhancing the primordial soup with minerals scraped off the land. This will help expedite the formation and evolution of life in Anillo's oceans. As on Earth, life will move onto the land and eventually animals will inhabit its continents. I posit that half a billion years after Noom arrives, "people" will first evolve onto Anillo. Because Anillo has a twelve-hour day when it captures Noom, the planet will be spinning much faster than the Earth does, so the cycles of life on Anillo will be different from ours.*

Likewise, the moon Anilloans see will be different from our Moon in several ways. After spiraling inward for half a billion years, Noom will be about $9/10$ as far away from Anillo as our Moon is from Earth. As a result, Noom will be orbiting faster, with a cycle of phases only $23\frac{1}{3}$ days long, rather than our Moon's $27\frac{1}{3}$ days. Noom will also appear about eleven percent wider and nearly twenty-five percent brighter than does our Moon. Being in the same orbit as our Moon, Noom will have solar and lunar eclipses, as discussed in Chapter 1. However, at the time when people first see it, Noom will not be so large that it undergoes eclipses every new and full moon, as occurs for Mynoa in orbit around Tyran (Chapter 2).

Before coming into orbit, Noom had a six-day rotation rate, which it got when it was locked in synchronous orbit with its pre-Anillo companion. It keeps rotating at that rate during the early years in orbit around Anillo because early on in their relationship, there is no way for the planet to change Noom's rotation rate. When people on the planet first see it, Noom rotates so that all sides face the planet at different times, unlike our Moon with its synchronous rotation.

---

* See the discussion in Chapter 2 about biological clocks.

That gets me to wondering: there is nothing in the sky visible to the naked eye that appears to be rotating on its own as seen from Earth. How might the fact that Noom's face keeps changing affect how early Anilloans relate to their moon compared to how our ancestors related to the Moon? For example, having the same side always facing Earth enhances the impression that the Moon orbits us. This, in turn, supported the belief that everything orbits the Earth, which was a common belief until a few centuries ago. Watching Noom spin throughout the ages would not help the argument of an Anillo-centered universe.

THE DEATH SPIRAL

It will take several billion years for Noom to reach its destiny. During that period, its elliptical orbit will cause the moon to heat, as occurred with Lluna in Chapter 1. Land tides will force Noom into synchronous rotation. From then on, the same half of that moon will always be visible from Anillo.

Anillo's Rotation

Anillo's rotation rate will also change under the influence of Noom. In fact, it undergoes an incredible transformation as Noom spirals inward. The planet's behavior is based on the conservation of angular momentum, discussed above.* Angular momentum depends on the direction something is spinning or orbiting. Suppose that you have two identical platters, like jugglers use, with one spinning clockwise and the other spinning counterclockwise at the same speed. The total angular momentum of the two, the sum of the two angular momenta, is zero. In other words, if you were able to spin them both up and then drop one platter on top of the other, they would both come to a complete stop. Conversely, if they are both spinning at the same rate in the same direction before you drop one platter on the other, then the total angular momentum of the two combined has twice the value that each platter has separately. As a result, when you drop one on top of the other, they will continue to spin together at the same rate.

---

* I am ignoring here small (often called "second-order") effects, such as the tidal force from the Sun, which are negligible compared to that of Noom as that moon spirals significantly inward.

Anillo and Noom have angular momenta in opposite directions. As Noom spirals inward its angular momentum decreases. That decrease is balanced by a decrease in Anillo's angular momentum in the opposite direction so that the combined total angular momentum of the two bodies remains constant. In graphic terms, as Noom spirals in and orbits faster and faster, Anillo's rotation rate decreases. The physical cause of Anillo's slowing is friction between its oceans and continents due to the tides that Noom creates on it.

When Noom is first trapped in orbit around Anillo, the net angular momentum of the system (angular momentum of Anillo minus the angular momentum of Noom) is in the direction of Noom's orbit. Equivalently, it is opposite to the direction of Anillo's spin. As Noom spirals inward, Anillo's rotation rate continues to slow until, when Noom has come in to one-quarter of the distance from which it started, Anillo stops rotating!

Noom is still orbiting, so that the total angular momentum of the system is still preserved. Furthermore, Noom still keeps raising tides on Anillo. At this point in their relationship, the tides on Anillo from Noom are sixty-four times higher than the tides were when Noom was captured or, equivalently, than the tides are on Earth today.

Because water has mass, it takes time for the high tides on Anillo to move in response to Noom's motion around the planet.* As a result, the high tide that Noom creates on the nonrotating Anillo closest to the moon lags behind Noom in its orbit. This tide therefore continues to pull against Noom as it orbits and so the moon continues to lose angular momentum and spiral inward. The kicker is that those tides continue to generate friction between the oceans and continents of Anillo and hence that world begins to rotate in the opposite direction from which it was spinning originally.

Climate Changes

During the millennia that Anillo's rotation is grinding to a halt and beginning to reverse direction, a day will roughly equal a year. At the equator during this epoch, the Sun will rise and move continuously across the sky

---

* A concept called inertia.

Figure 3.3: Slow-motion rise of Anillo's Sun as seen at equator. CREDIT: W.H. FREEMAN & CO.

for up to 182.6 (present) days before setting (Figure 3.3). It will be night for the same length of time. Depending on the tilt of Anillo's axis during this period of its life, the length of light and darkness at some latitudes will be much longer or shorter than this, as we see on Earth above the Arctic Circle and below the Antarctic Circle.

Half an Earth year of continuous daylight on Anillo will cause tremendous heating of the continents and oceans. This heat, in turn, will generate powerful hurricanes that could last for weeks or longer and cover large fractions of the daylight side of the planet. At the same time, the heat will cause much more evaporation of water than occurs during a day on Earth. By sunset in most places on Anillo during this era, the climate will be tropically hot and humid.

Imagine a palm tree that sprouts at sunrise during this epoch on Anillo. Continually receiving sunshine for six Earth months, it would grow steadily, while water evaporating from the oceans will increase the humidity, making it even more favorable for vegetation. Come sunset, the heating stops, snow begins falling, and the land and air around our palm tree begin to cool. In fact, they cool continually for another six Earth months, reaching temperatures of below −125°F. There is no way that our palm, or any other animal or vegetable life accustomed to tropical climates, is going to survive. Indeed, it is likely that during this period of Anillo's existence most cold-blooded creatures will become extinct.

During the night of this epoch, most lakes will freeze to great depths. Ice floes will grow equatorward from the polar regions, but the oceans will not develop permanent ice covers. In the first place, the surface areas of the oceans are so large that there isn't enough time in six Earth months for them to become completely covered with ice. In the second place, as the ice sheets become larger and larger, they are continually being broken up by the tremendous tides that Anillo will then be experiencing.

Endgame

As Anillo spins faster and faster in the opposite direction from which it had been going originally, the day will become more "normal." By the time Noom has come inward to $\frac{1}{10}$ its original distance from Anillo, the planet will be rotating about once every twenty-four hours. The Sun will rise in the west and set in the east, and the tides on the planet will be 1,000 times higher than the tides we experience on Earth today. Clearly the concept of a shoreline will be profoundly different than what we think of today. Here in Maine, tides are often ten feet high; tides on Anillo with Noom that close would run two miles high.

Returning to the question of whether Noom spirals all the way in and collides with Anillo, we need to revisit the two key distances from a planet related to where moons can exist. Recall from Chapter 2 that synchronous orbit is the distance of the moon's orbit at which the planet rotates (spins) at the same rate that the moon revolves (orbits). A moon initially in a retrograde orbit, such as Noom, does not enter the synchronous orbit of its planet either when the planet is rotating in its initial direction or after it reverses direction.

Noom will, however, encounter the other key distance, Anillo's Roche limit. This is the same Roche limit that prevents rings inside that radius from becoming moons. If a moon held together by its gravity (rather than just being a big rock in space) ventures inside the Roche limit, the gravitational tidal pull from the planet will exceed the gravitational pull that the surface of the moon feels from the rest of the moon. The side of the moon facing the planet will be lifted right off the moon, a process that will continue until the moon is completely disassembled.

Noom's fate is to be pulled apart. By the time Noom orbits down to the Roche limit, it will be in synchronous rotation around Anillo, so the same side of the moon will continuously face the planet, making it easy for the planet to peel the moon apart. Instead of spiraling onto Anillo, Noom is destined to become a giant ring of debris orbiting about 2,800 miles above the planet. It is extremely likely that collisions among pieces of debris would cause a more or less continuous rain of pebble- to boulder-sized particles entering Anillo's atmosphere, especially in the equatorial latitudes, above which Noom orbits. Although many pieces of this debris would burn up on the way down, the surface there is likely to get a good pummeling.

Once the ring has formed, the tides on Anillo will drop precipitously to about one-third what they are on Earth today. These would have a small effect on the ring, causing parts of it to bunch together and then move apart as the debris, orbiting faster than the planet rotates, zips by the regions of high tide.

Life?

The process of Anillo's rotation slowing, stopping, and speeding up in the opposite direction occurs over billions of years. Therefore, it is likely that life on Anillo could adapt to the changes in that world's rotation rate. Recalling the discussion in Chapter 2, one major evolutionary change that would have to occur is the change in the rate at which biological clocks operate. Going from a ten-hour day to a day when the Sun never sets and then to a twenty-four-hour day would require considerable biological flexibility.

Given the extraordinary change in temperatures on Anillo throughout the year when its day is half a year long, I expect that animal life there would mostly evolve into migratory species, moving perhaps several times a year to places where the climate is most consistent with its capabilities. When the day becomes more "normal," it is likely that animal life would evolve again, with many species losing the need for migration.

What If the Earth's Crust Were Thicker?

DICHRON

4

NOBLE LECTURE, 1933

It is the greatest honor of my life to share the thirty-third Noble Prize in Physics. I realize that you have bent the rules in order to give it to me. I guess it shows flexibility on both of our parts. I hope this will become clear as I relate the story of my discovery. I am, as you know, a physician by training. I was ship's doctor on the MS *Borealis* on a cruise to explore the north polar region of the planet twenty-six years ago. We had been sent to try to help figure out the cause of the global warming that our planet Dichron is presently undergoing.

Our expedition was initiated after Eskimo fishermen reported seeing unusual cloud formations and exceptionally mild weather in the northern reaches of the continent. The men were reporting temperatures in the fifties, where they should have been below zero.

We took on coal at Newcastle and steamed north. That was my first trip above the Arctic Circle, but others on the crew were dumbfounded by the weather we encountered. At latitudes where there historically had been icing on the hull and rigging, we had balmy days and tepid nights. There was a mix of clouds and sunshine at first. As we got farther north, we encountered full overcast with increasingly heavy showers. But they were warm showers, the kind you experience near the equator. On previous cruises, I was told, precipitation would have been ice and snow, making it utterly impossible to walk on deck without studded boots and safety harnesses. For those of us without bathing suits, we were out, if you will excuse the expression, in our skivvies.

As you all know, the North Pole is normally capped with ice that spans over 500 miles and is a dozen feet thick. I can tell you categorically that there is no ice in the Arctic today. We steamed over the top of the world. Too bad it was cloudy, as the scientists on board wanted to stop at the Pole and watch the stars turn around them.

You may be wondering why there has been so little rise in the ocean levels with all that ice gone. It turns out that ice floating on liquid water pushes down exactly as much as does the same mass of liquid water added to the existing liquid water. So, when the ice melted, the water levels didn't change. It is only when ice on land melts and runs into the ocean that the water level changes.

Besides clouds, rain, and extraordinary temperatures, we found nothing unusual until we were 350 miles beyond the Pole from here. It was there that the entire ship's crew, without exception, developed the following symptoms: irritability, claustrophobia, acute seasickness, and for a lack of a better descriptor, an overwhelming desire to be any place except where we were. I documented all this, interviewing as many crew members as possible.

We were entering an area where fog, often smelling like rotten eggs, covered the entire surface of the ocean. Indeed, in places the water seemed to simmer, like soup on the stove. The men were by this time mutinous, but our captain, Captain Deming, was able to maintain order, if not to say discipline. He decided to send a man down in a diving bell to take temperature measurements. This turned out to be a mistake, as the water was so hot that the diver in the bell was scalded by the time he reached ninety feet.

A geologist floated the idea that Dichron is sufficiently hot underground that rock there is molten, and that the planet may have flaws in its crust which allow some of this liquid rock to escape. Using that hypothesis, the team of geologists on board concluded that an undersea volcano was erupting below us. Volcanoes are hypothetical tubes of molten rock that emerge from deep inside our planet. Of course, we have never seen an active one on land, but mountains such as Etna and Fuji, with depressions at their peaks, may well be extinct volcanoes. Captain Deming decided to steam on so that we could get horizontal temperature measurements across the region of bubbling water. Those data would give us an idea of how wide is the suspected volcano. The temperature did not drop for the next 250 miles. We changed direction and made several passes to map out the entire overheated region. The uniformly hot region is roughly circular and has a diameter of a little over 300 miles.

During all of this time, the men were absolutely on the edge of control and, if I may say, on the edge of sanity. They really wanted to leave the Arctic. I confess that I did, too. Only Captain Deming's iron will kept us there. He later told me, over a pint, that it was the hardest time of his life.

As we sailed home, I spent countless hours discussing the results of our observations with the scientists on board. They could not reconcile the huge size of the so-called volcano with any models of volcanic activity. I, on the other hand, could not explain the irrational behavior of the crew.

The answer came to me in a dream. Or at least the crucial connection between the men and the sea did. I dreamed that I was on a roller coaster, which was scary, but exciting, when suddenly the entire frame of the ride began shaking, as if there were pistons underneath it or as if it were beginning to fail structurally. The excitement turned to terror. I wanted to be anywhere else. Then I awoke. What if our responses to what was happening below the Arctic Ocean were built into our psyches? Maybe it is a protective device that we got from our ancestors who had to escape when similar events occurred in the past, possibly even on one of the continents.

I presented these ideas to the scientists at our next meeting. That led the geologists to develop the idea of magma extrusions that bring titanic volumes of molten rock to Dichron's surface, thereby relieving the planet's internal stress. Further observations and studies have shown that this does indeed occur. Indeed, it has been suggested that large melts may even occur on dry land. This theory, as you know, is causing no end of social upheaval.

The surface of the Earth has undergone episodes during which vast volumes of magma flowed out and covered thousands of square miles of our planet's surface. The Columbia plateau, covering over 60,000 square miles of Washington, Oregon, and Idaho, is one example. It formed over an eleven-million-year period ending six million years ago. The Siberian Traps, another such region covering over half a million square miles of Siberia, were formed about 251 million years ago. Some of Earth's mass extinctions, such as the Permian–Triassic extinction, which also occurred 251 million years ago, may well have been byproducts of some of these "flood basalt" events.

Although they caused some of the most widespread changes to the Earth's surface in relatively recent geological time, flood basalts still represent leakage of molten rock from only a few localized sources. In this chapter, we explore the situation in which every place on a planet's surface faces such events from time to time. The cause of this different surface behavior on the new planet, which I call Dichron, is that it has a much thicker crust than the Earth. This may sound like it would be a good thing, preventing such unfortunate events as earthquakes and the resulting tsunamis and related disasters, but these advantages come at a horrific price.

Our planet's interior is hell in Earth, full of rocks, metal, and gas, some of which are hotter than the surface of the Sun (5,800 K/10,000°F). Fortunately for us, some of the Earth's interior heat leaks out on a regular basis through the crust and through volcanoes and lava seepage on the boundaries between tectonic plates, which are discussed shortly. In this chapter we explore the consequences of keeping hell pent up inside the planet Dichron, which has the same mass as Earth and a single moon called Vault. Dichron orbits the Sun at the same distance as does Earth.

Chapter 1 briefly introduced the scenario by which astronomers believe Earth formed in a swirling disk of gas and dust. To understand Dichron and how it differs from our planet, it is worthwhile exploring planet formation in more detail. I focus on the Earth here and then consider what we need to do to morph it into Dichron. Recall that stars and planets form from the gravitational collapse of tiny fragments of giant interstellar clouds.

The Sun formed in the center of a collapsing fragment. That fragment, like the swirling wisps of smoke rising from a campfire, was rotating. As the fragment collapsed, an orbiting disk of gas and dust formed around the young sun.

At first, the gas and dust particles collided randomly (their mutual gravitational forces were too weak to pull them together). Depending on what chemicals the particles contained, they bonded together, repelled, or ignored each other. Eventually, when pieces of debris became rock- and boulder-sized, their mutual gravitational attraction enhanced the process of bringing them together.

If two pieces of rubble are colliding sufficiently slowly, they will merge into one larger piece. If they collide too quickly, they will smash each other to smithereens and the process of creating larger pieces of debris (which can eventually become planets or other large orbiting bodies) will be temporarily sidetracked for them. Impacts of intermediate intensity can cause larger pieces of debris to break into a few pieces that orbit each other.

After a few million years, the collisions that created larger bodies prevailed over the collisions that destroyed bodies and as a result, myriad planetesimals, each at least a mile across, formed. Collisions of these bodies led to larger "protoplanets" at least the size of our Moon. Protoplanets are bodies that contain enough matter to ensure that further collisions with planetesimals will cause them to grow. As they grow, protoplanets cleared the regions in which they orbited by pulling the remaining nearby debris onto themselves or into orbit around them, thus becoming planets (with moons, when some debris went into orbit).

The debris that formed the Earth varied in composition, including small amounts of hydrogen and helium gas, other compounds including carbon dioxide, methane, ammonia, nitrogen, and water, as well as dust particles rich in silicon, aluminum, iron, calcium, sodium, potassium, magnesium, and other elements.

EVOLUTION OF EARTH'S INTERIOR

The interior of the newborn Earth was heated by at least three major effects (other sources are introduced later in the book). First, the innumerable impacts that occurred over the course of hundreds of millions of years sent shock waves deep into the planet. These waves compressed the interior through which they traveled, meaning that they increased the pressure down there. Whenever anything is compressed, it heats up. In this case, that heating was substantial. To get a feel for the scale of such events, the impact of a piece of space debris 90 yards across moving at 30,000 miles per hour (typical sizes and speeds during the period Earth was forming) would generate an explosion equivalent to that of 50 million tons of TNT exploding all at once.* The region below such an impact would be compressed by a mile. A column of rock extending downward for miles would have instantly been turned molten. At the same time, millions of cubic yards of the planet's surface would be vaporized, shoot upward, and then fall back down. The craters created by such impacts would each be several miles across.

The second major source of heat inside the Earth was the decay of radioactive elements that were incorporated in it. Radioactive elements are those whose nuclei spontaneously break apart, or fission. Such events are accompanied by the release of energy as the pieces of the nuclei fly apart and high-energy radiation, such as x-rays, is emitted. All of this energy is absorbed in the surrounding rock, heating it up.

The third major heat source, related to the first one, was the compression of the inside of the planet by the gravitational force that the young Earth exerted on itself. Imagine a bag of garbage, which I name the alpha bag, in a garbage dump. Along comes a garbage truck, which dumps its load on top of the alpha bag, thereby crushing it. Another truck dumps its load on top of the first load, crushing the alpha bag even further. At the same time the bag and its contents are becoming flatter, they are also heating up, which you could prove by having a thermometer inside the bag. Analogously, as more and more debris rained down on the young Earth, its mass grew, along with the pressure on its interior. This helped heat its insides.

By the time the Earth was just ten million years old its insides were sufficiently hot to ensure that all the matter down there was molten. This enabled the denser elements, especially iron and nickel, to sink to the center of the planet. They descended for just the same reason that a rock sinks when

---

* This is the energy that would be released by the most powerful thermonuclear weapon about which I could find information. Such an explosion would destroy virtually any city on Earth.

thrown into water. Conversely, lighter minerals, rich in silicon, aluminum, calcium, and other lighter elements and compounds, especially oxygen-rich ones, floated upward.

Without this process, called differentiation, occurring, an Earthlike planet would have a surface containing many more iron-rich compounds than does Earth. Indeed, a third of the Earth's mass is composed of iron, virtually all of which is in its cores. On an undifferentiated planet, a third of its surface would be iron compounds. In comparison, only five percent of the mass of the Earth's crust is iron. As we now explore, interactions between iron and oxygen on an undifferentiated planet would have led to an oxygen-poor atmosphere, drastically decreasing the prospects for animal life there.

The interplay between iron and oxygen stems from the fact that oxygen reacts extensively with a variety of elements and compounds. For example, the interaction between oxygen and hydrogen is so energetic that NASA makes rocket engines based just on this reaction. The first sustained flow of oxygen into Earth's atmosphere $3\frac{1}{2}$ billion years ago came from ocean plant life, which took in carbon dioxide and gave off oxygen as a waste product of photosynthesis, as discussed in Chapter 2. The oxygen that didn't remain dissolved in the oceans escaped into the air, but it didn't stay there. Rather, that early atmospheric oxygen combined with many minerals on the Earth's surface. What little iron there was on the surface was quickly combined with oxygen to create iron oxides, which we know commonly as rust. It was not until all the minerals on the Earth's surface capable of combining with oxygen had done so that this gas was able to remain in the atmosphere and provide the energy required for animal life here.

On an undifferentiated planet, with a third of its young surface composed of iron, the process of locking oxygen with that iron would not yet be over. Even worse, after all the surface iron and other elements were sated with it, there might be insufficient oxygen left over to provide enough oxygen in the air for animals like us to exist on that planet. But no animal life means no plant life: the plant life that had been using the carbon dioxide in the early atmosphere to grow would have used it all up and there would be no new carbon dioxide (from animals) for them to use in photosynthesis. The plants would die and the planet would become essentially lifeless, at least as far as life as we know it is concerned.

So differentiation was a good thing from our perspective. It put most of the Earth's iron out of reach of the oxygen gas that started pouring into the atmosphere from the oceans. Happily for the prospects of life there, Dichron also differentiated.

The other major event that is extremely important in understanding the difference between Earth and Dichron, with its thicker crust, is the formation of a planet's solid crust.

Earth's Interior

Having shown why the outer layers of Earth and Dichron are composed of lighter minerals than the interior, we now consider the process by which their crusts came into existence. Throughout the first 100 million years or so of their existence, the heat from inside them and from impacts on the young planets was so intense that their surfaces remained essentially molten. Only after the initial periods of bombardment were over did the young planets' surfaces become cool enough for a serious, long-lived crust to begin to solidify. Let's focus on Earth, first.

The Earth's crust is by no means uniform. The lower-density rock comprises the higher-rising land masses such as continents, whereas the higher-density, more compact rock creates the ocean bottoms. It is through this crust that heat from inside the planet escapes.

Three major factors—temperature, pressure, and chemical composition—determine whether the various layers of the Earth are solid or liquid. Starting on the surface, the Earth's solid crust ranges in thickness from less than six miles (under the oceans) to over forty miles in mountainous regions of the continents. The chemistry of the rock changes below the crust, bringing us into the mantle. The top layer of the mantle, extending down to a depth of about sixty miles, is solid. It and the crust combine to create a layer called the lithosphere (Figure 4.1) composed of six large and numerous smaller regions called tectonic plates that are floating on the pliable rock below them.

The mantle rock below the lithosphere extending down to about 200 miles is so hot that it is plastic, deforming under pressure like Silly Putty. This region is called the asthenosphere. Below the asthenosphere, the pressure is so high that despite temperatures of over 1,000°F, the rock down there is nearly solid. This region, sometimes called the deep mantle, descends down to 1,800 miles below the surface to the realm of the iron core. It is important to note that the

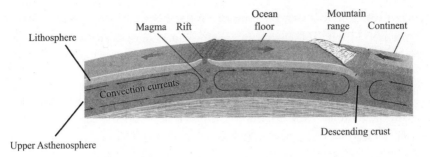

Figure 4.1: *Earth's upper layers: simplified drawing of the lithosphere and upper astheno-sphere.* CREDIT: W.H. FREEMAN & CO.

temperature and pressure combine to create an outer, liquid iron core and an inner, solid iron core.

Conveying Heat Out

Heat flow inside a planet is very complicated. Geologists are still actively working to create models that accurately replicate what is happening inside the Earth. We do not need to explore the details for our purposes. The basic model of heat transfer out of the Earth is similar to the process that causes soup to simmer. Heated by the liquid core, blobs of the bottom of the rock layer that rest against it expand (most things expand when heated). When a volume expands, it is less dense than the surrounding rock and so that blob begins to rise: it is buoyant, like a stick you hold under water and then release.

The blob of heated rock ascends until it reaches the bottom of the astheno-sphere, which it then heats. By doing so, the blob cools and it is shoved side-ways, out of the way of other rising, hot blobs of slowly oozing rock. As a result of giving off its heat to the asthenosphere, the initial blob becomes denser than the surrounding rock (most cooling things become denser), and begins to descend back into the asthenosphere. This cycle of heating the mantle from below, followed by the hot lower-density rock rising, followed by cooling of the rock as its heat is transferred to the asthenosphere above it, fol-lowed by descending of the cooler higher-density rock, is called convection.

Once heated, the asthenosphere repeats the cycle of convection, carrying hot rock farther upward to the lithosphere. Just as you feel less pressure when rising up in water, the pressure on the molten asthenosphere rock decreases as it rises. This change in pressure causes some of the rock to change from flowing slowly, like thick porridge, to a much more liquid state, like syrup.

The convective flow of the asthenosphere brings moving rock into contact with the bottom of the lithosphere, which it heats. As the cooling asthenosphere rock moves sideways in the convection cycle (Figure 4.1), it causes the top layer of the Earth, the lithosphere, to move with it. This creates the flow of tectonic plates. At the same time, heat from the asthenosphere creates blobs of molten rock in the lithosphere. This hot rock rises up and emerges as lava at the boundaries of the plates (in the ocean bottoms) as well as through volcanoes.

Carried upward in the form of molten rock, the heat from inside Earth leaks out primarily in two kinds of places: the boundaries where tectonic plates separate (see Figure 4.1) and volcanoes. Consider, for example, the separation of the Eurasian Plate (carrying Europe and Asia) and the North American Plate. They move apart along a line running roughly north–south down the middle of the Atlantic Ocean called the Mid-Atlantic Ridge. In that region of the crust, molten rock is continuously pushing up and out, warming the ocean bottom from unheated temperatures of 33°F to temperatures of 750°F, and thereby removing heat from inside the Earth. This heat is carried upward in the ocean currents and eventually is radiated into space.

Volcanoes

Heat-emitting volcanoes come in a variety of flavors. Let me mention just two: stratovolcanoes and shield volcanoes. Stratovolcanoes are those that occur with massive blasts and plumes of gas and debris that rise high into the air. These often release heat and other energy from inside the planet in a matter of a few days, although some persist for weeks, months, years, or even millennia. Mount Stromboli, off the north coast of Sicily, has been active for over 2,000 years. Other stratovolcanoes, of which there are over 700, include Mount Saint Helens in Washington state, Mount Vesuvius in Italy, and Mount Fuji in Japan.

Shield volcanoes emit lava at a more leisurely rate and can continue to do so for hundreds of thousands of years. The Hawaiian Islands were (and continue to be) created by shield volcanoes. They do not explode. Rather, lava seeps out of them.

There are typically 20 active volcanoes on Earth at any time. Between 50 and 70 erupt each year. In the past 10,000 years some 1,500 different volcanoes have been active. The total amount of power continuously emitted by all the volcanoes on Earth amounts to over 1.6 billion watts (compare this to the heat given off by a typical 100-watt light bulb). Moreover, the power emitted by the mid-ocean ridges and various other heat vents in the Earth's surface is staggering, over 25,000 times greater still. All of this energy, plus other energy in the form of earthquakes, comes from inside the Earth, meaning that our planet continually gives off over 40 trillion watts, an incredible amount of energy!

This process of generating and giving off energy from inside the Earth has been going on since the planet formed, and it has been pretty much in equilibrium, meaning that as much heat leaves the surface as is generated inside, for over three billion years.

FORMATION OF EARTH'S ATMOSPHERE

Before morphing from the Earth's crust to Dichron's thicker one, we need to be aware of two other things about Dichron that could also be different from the Earth: Dichron's atmosphere and its oceans. The air we breathe is utterly unlike the early atmospheres that the Earth had. Our air represents the third atmosphere that has surrounded this planet. The first one was composed primarily of hydrogen and helium gas that remained after the young Sun had removed most of these elements from the inner solar system, as introduced in Chapter 2. It didn't last long.

Gases are attracted by the gravitational force from their world. Whether this force is great enough to keep them in the atmosphere depends on how fast the gas molecules move. This motion is a measure of their temperature. If a gas is hot enough (i.e., moving fast enough), it can escape from the gravitational force holding it down. In that case we say that the gas has "escape velocity." Such gas drifts into interplanetary space, never to return.

The lighter a gas, meaning the lower the mass it has, the easier it is for that gas to be given enough energy to achieve escape velocity. That energy comes from any heat sources the gas encounters as well as from collisions between gas particles in the air and gas particles coming in from space. Hydrogen and helium have the lowest masses of any elements and as gases they are the easiest to speed up to escape velocity. When the Earth's atmosphere was composed of these elements, heat from volcanoes, lava flows, and the Sun, along with impacts from particles flowing out of the Sun, called the solar wind, quickly energized that first atmosphere and caused it to evaporate into space.

The Earth's second atmosphere came primarily from inside the planet and was vented out through volcanoes and the boundaries between tectonic plates. This gas was composed primarily of carbon dioxide, along with a small amount of nitrogen and traces of other gases such as argon. This second atmosphere, about ninety times as dense as the air we breathe, had virtually no free oxygen molecules, which are essential for animal life today. As discussed in Chapter 2 and earlier in this chapter, this atmosphere was transformed into the third, nitrogen-oxygen, atmosphere we breathe with the help of the oceans and plant life.

It was easier for Earth to retain these latter two atmospheres than the original one. The gases in them, such as carbon dioxide, oxygen, nitrogen, and argon, each have so much mass that they cannot be easily energized to escape velocity. Therefore, the gases in the second and third atmospheres each remain around Earth for billions of years. This brings us to the issue of how the oceans formed. Recall that in Chapter 2 I explained that most of the Earth's water was delivered by impacts of asteroids (rock, metal, and ice-rich bodies), along with impacts of comets (rock and ice-rich bodies). Probably the largest impact that the Earth experienced was the one that splashed into orbit the debris that formed the Moon. This event may also have deposited a huge volume of water here. Over tens, perhaps hundreds of millions of years, water from inside and outside the Earth accumulated on its surface, creating the oceans.

The formation and evolution of the Earth's interior, crust, atmosphere, and oceans were all essential ingredients in the formation and evolution of life here. We are now ready to ask the question: could Dichron, Earth with a thicker crust, develop a surface, atmosphere, and oceans necessary for complex life to evolve? The short answer is yes. The longer answer is that the life there would face some very different challenges than does life on Earth.

DICHRON AND ITS THICKER CRUST

The key difference between the lithosphere (crust and upper mantle) of Earth and the lithosphere of Dichron is the amount of water each contains. Unlike Earth, I posit that Dichron developed its lithosphere before there was significant water in those rocks. Water acts as a lubricant in the Earth's lithosphere, making this region sufficiently flexible so that the convection of the asthenosphere below it can force the lithosphere to slide. As discussed earlier, this motion creates the tectonic plates, along whose boundaries lie most of the volcanoes on Earth. Because Dichron was created with much less water in its lithosphere, that world developed a completely inflexible outer layer. To make this happen, I set Dichron to forming in an environment in which water-rich debris falls on it later than occurred on Earth.

When Dichron's lithosphere was young and thin, blobs of magma that were especially large and especially hot melted their way straight up through it, as also happened on Earth. As Dichron aged, its lithosphere became more rigid than did the Earth's lithosphere because there is little water inside Dichron. This unyielding lithosphere is extremely resistant to sideways motion because as one part of it is being shoved sideways (by the asthenosphere below it), the rest cannot get out of its way.

Unlike the Earth, the convective motion of Dichron's asthenosphere and the blobs of liquid magma that form in it are unable to pierce most of the lithosphere. Most of the asthenosphere that reaches the bottom of the lithosphere just heats it, thereby cooling itself off. That heat transferred to the rocky lithosphere slowly moves up through it to the surface. From there it heats the atmosphere, which in turn radiates this heat into space. This process also occurs on Earth where the lithosphere is not broken by rifts or volcanoes. The catch is that with a thick lithosphere, hereafter meaning thick like Dichron, the heat escaping through it is less than occurs on Earth.

Whereas local regions of the Earth's surface are extremely hot (i.e., volcanoes and mid-ocean rifts), the entire crust of Dichron is heated more uniformly. Therefore its overall surface temperature is a few degrees warmer than that of Earth, but without nearly as many hot spots. Because it remains fixed in place longer under Dichron's lithosphere, more of the cooling asthenosphere rock ends up solidifying and becoming part of the lithosphere than does the rising rock inside Earth. As a result, Dichron's lithosphere thickens, and Earth's lithosphere maintains roughly the same thickness for billions of years.

The few volcanoes that punch through Dichron's young lithosphere remain fixed in location on the planet's surface, unlike many of the volcanoes on Earth, such as those that are forming the Hawaiian Islands. These latter volcanoes are changing location as the Pacific Plate slides northward. In the case of those islands, volcanoes become dormant, islands weather away and disappear beneath the waves, and new volcanic islands form in different places. Unless a volcano on Dichron stays active, the thickening lithosphere below and adjacent to it will cause it to seal up and become extinct. And, as the lithosphere of the planet thickens, it will become harder and harder for blobs of magma to punch through it and form new volcanoes.

That Thicker Lithosphere

Dichron's thicker lithosphere is the key to the difference between that planet and the Earth. Because tectonic plate motion is stifled on Dichron, the cooling of its interior is slowed compared to what occurs inside the Earth. The rub is that the heat created in Dichron's interior doesn't just vanish. The law of conservation of energy* that we are taught in science classes requires the heat generated inside a planet to either stay there or get out. Therefore, the heat under Dichron's lithosphere, brought from the core, builds up; the asthenosphere gets hotter and hotter because it can't cool efficiently until something gives.

Insight into the fate of the heat in Dichron comes from the work of planetary geologists who had to explain the evolution of the planet Venus, which also lacks plate tectonics. The heat stored under Dichron's lithosphere is not distributed uniformly. Where convection is bringing up more heat from deep inside, the bottom of the lithosphere gets especially hot. When a local region of the bottom of Dichron's lithosphere gets sufficiently hot, the rock there melts. This is a runaway process, meaning that melting continues all the way to the surface. Every few tens of millions of years, each tract of land under Dichron's oceans or on its continents becomes molten. As a result, the

---

* Although true for situations such as we are discussing here, this law is not true deep inside stars and other places where mass can be converted to energy or energy to mass.

heat from inside has an escape route, wherein that part of the planet's interior cools enough to let the lithosphere resolidify. Then the process of thickening, heating, and melting begins again.

Based on models of Venus's lithosphere, I expect that each region of Dichron's surface roughly 250 miles across melts every twenty-five million years. The molten surface allows large amounts of heat to leave the interior. Within a few hundred years the new surface solidifies. Recall from the beginning of this chapter that analogous events have occurred on the Earth, but not nearly to the extent they occur on Dichron. Meltdowns on the continents there will erase all features and all life in those regions, replacing the surface there with a new lithosphere that may have a very different shape or topology than the surface it replaces. Likewise, the new surface may have a very different chemistry than did the previous one. Furthermore, when a region of the surface is molten, lava from it will flow downhill to fill in adjacent areas, meaning that the "damage" each meltdown causes may extend beyond its boundary.

Molten surfaces aren't the only consequence of these events. Huge amounts of gas are also released from the interior. Containing water vapor, carbon dioxide, hydrogen sulfide (which smells like rotten eggs), and sulfur dioxide, this gas will affect the atmosphere by creating large quantities of sulfuric acid, among other nasty compounds. Newly melted areas will be vast wastelands, probably for centuries, as gases continue to leak out of them and their surfaces cool and solidify. Because a meltdown will be fatal for all life in each such region, I expect that the evolutionary process on Dichron will lead to sensory equipment in the animal life, warning it to flee when, for example, the surface temperature begins to rapidly rise or when there are certain gases in the air.

Meltdowns on the ocean bottoms will lead to a series of significant climatic changes. First, large volumes of ocean water will be heated dramatically and some of it will be vaporized. Depending on the breadth of the melt, some or much of that water will rise to the air as water vapor, increasing the humidity of the air and temporarily causing the ocean levels to drop. Gases such as carbon dioxide, dissolved in the water, will be released into the air. The heated ocean water will cause the formation of hurricanes (typhoons), as heated water does here on Earth. It is likely that the enormously greater heating of the water on Dichron by the underwater melts (compared to any heating of oceans on Earth) will create much stronger, longer-lived storms than we ever experience here.

The tremendous increase in moisture in the air each time the ocean bottom melts will also lead to a very complicated change in surface temperature. This will be caused in part by the water vapor in the air absorbing heat from the Sun and from the surface, thereby warming the air more than usual. The extra clouds created by the moisture will, on the other hand, decrease the amount of direct heating of the land by sunlight. It is likely that the moisture lost by the oceans due to melts will return as rain, but the process of heating and vaporizing ocean water will lead to very complicated weather patterns that will present challenges to everything living on Dichron. Of course, once evolution has provided enough early warning and "flight" response tools to animals there, I believe that they will be able to take catastrophic changes to their world in stride, changes that would stagger our imaginations if they occurred on Earth.

Dichron and Its Land, Water, and Air

With the picture of epic changes to its surface now in hand, we can put the finishing touches on Dichron and step back to appreciate how life would evolve on this tectonic plate–free world. One crucial factor is Dichron's distance from its star. The planet orbits it at the same distance we orbit the Sun. Like Earth, Dichron was struck by a Mars-sized body that tilted it over, creating the seasons, and splashing debris into orbit to form its moon, Vault. However, this impactor contained much less water than did the one that struck Earth. As a result, young Dichron had much less water introduced into its mantle and crust (helping to justify the lack of tectonic plates).

The formation of Dichron's original crust also paralleled that of the Earth. That is not to say they are identical. Although the surface of Dichron differs from the Earth in details, they both have higher lands (called continents on Earth and highlands on Venus) and lower lands (ocean basins on Earth and lowlands on Venus). To compare, Earth's continents cover about thirty percent of its surface, whereas highlands on Venus cover about ten percent of its surface; I give Dichron three continents, each covering only five percent of its surface. One lies in the northern hemisphere and the other two straddle the equator on opposite sides of the planet.

The three continents, separated from one another by roughly 9,000 miles, remain in those locations as long as they exist on the tectonic plate–free surface. The two equatorial continents are profoundly different from each other. One, Sluria, is a lush land of rolling plains, with a few mountains. The other, Craag, is mountainous, broken by deep verdant valleys that meander down to the oceans. Many fewer volcanoes than exist on Earth dot Dichron's surface. Most are extinct, strangled from below by the thickening crust. The longer-lived ones have created some of the highest mountains on that world.

The remaining eighty-five percent of Dichron is covered with lowlands that fill with water to become oceans. However, the ocean bottoms are decidedly different from those on Earth. Instead of having continental shelves and deep mid-ocean trenches, such as are found in the Atlantic Ocean, the land extending from the continents of Dichron descends gradually, over thousands of miles, into moderately deep, flat regions.

Dichron's atmosphere evolved similarly to ours, although the amount of gas eventually released from Dichron was lower. Like Earth, its first atmosphere was hydrogen and helium, which quickly evaporated. As noted earlier, through the molten surfaces of young planets rise vast quantities of carbon dioxide, water, and some nitrogen. As their surfaces solidify, this outgassing decreases, but some of it continues through volcanoes and other openings. Dichron's second atmosphere, vented out of that planet through fewer volcanoes and no rift valleys, was thinner than ours. It has only seventy-five percent of the gas as our present atmosphere. As discussed earlier, this gas was converted into the third, nitrogen-oxygen atmosphere. Recall that this transformation begins with algae in the oceans and, when surface life evolves there, the conversion is complemented by the transformation of carbon dioxide into oxygen by surface plants.

Because Dichron has fewer and smaller continents than Earth, it is a good thing that most of Dichron's land is near the equator, where it experiences a lot of rain and is mostly jungle. This dense foliage generates nearly as much oxygen from carbon dioxide as all the plant life on Earth. Allowing that Dichron's second atmosphere is less dense than ours, its final oxygen-nitrogen atmosphere ends up being about eighty-five percent as rich in oxygen as the air you breathe. Air at sea level on Dichron is therefore as dense as air at the altitude of one mile on Earth. If you are reading this in Denver, at an altitude of one mile, you know what the sea-level air is like on this new world. All other things being equal,* when "people" evolve onto Dichron, it is likely that they will have to have bigger, more complex lungs and bigger chests in order to draw in the same amount of oxygen as we take in today.

The Recession of Vault

Fixing the locations of Dichron's continents changes the history of the moon Vault, compared to that of our Moon. We saw in the previous chapters how the tides generated by a moon act back on it. If, as with our Moon or Vault, a moon orbits in the direction the planet rotates, the tides cause it to spiral away from its planet. One of the key factors that determines how quickly a moon recedes is how much resistance the tidal waters in the oceans feel from the ocean bottoms and from the continents they encounter. As the planet turns and the moon orbits, the high tide nearest it is flowing in an effort to be right under the moon. If the planet were rotating slowly enough and continents didn't get in the way of the water flow, then the tide could be directly between the centers of the planet and moon. If the high tide were to remain directly between the centers of the two bodies, the moon would not be pulled forward by it, so the moon would remain in the same orbit forever. The more that the high tide is pulled away from that location by the planet's rotation, the more the tide pulls the moon ahead in its orbit, causing it to spiral away.

Both our Moon and Vault spiral away from their planets because Earth and Dichron are spinning too fast to allow the water to remain directly below the moons. Given the locations of the continents on Dichron and the fact that the ocean bottoms on that world are smoother than those on Earth, how much will the tides on Dichron pull on Vault and cause it to spiral away compared to the same effect by the Earth's tides on our Moon? The ocean tides on Dichron encounter less resistance from changing ocean bottoms and the relatively small continents, which resemble Australia much more than the Americas or the Eurasian landmass.

---

* Which, of course, they never are.

The shapes of the continents affect the recession of a moon. Here on Earth the continents run from over 70° north latitude all the way down to 55° south latitude, breaking up the flow of ocean tides and thereby preventing a high tide from remaining near the Moon; the tides on Dichron will flow around the continents much more smoothly. As a result, the high tide closest to Vault will not be ten degrees ahead of that moon, as the equivalent tide is here on Earth, but more nearly aligned between the centers of Dichron and Vault. In other words, because Vault is pulled forward less by the tides on Dichron than our Moon is by the Earth, Vault will spiral away more slowly than our Moon.

As a result of the slower recession, Dichron will not slow down as quickly as the Earth. Therefore, when people evolve there, the day will be shorter on Dichron than our twenty-four-hour day. At that time, a day on Dichron will be about twenty hours long, and Vault will only be three-quarters as far away as our Moon is from Earth. The difference in the length of the day will force a fundamental difference in the evolution of life on Dichron than on Earth, namely that the biological clocks will run at a different rate there.

Biological Clocks

Humans would not enjoy living on Dichron. Expanding on the issue of biological clocks introduced in Chapter 2, suppose that astronomers locate Dichron and you volunteer to go live there. Suppose further that the technology exists to get you there before you die. How well would you fare? Unfortunately, the difference in the length of the day between Earth and Dichron is a showstopper. As we saw earlier, biological clocks help reset our bodies for cyclic activities. Biological clocks on Earth evolved to function in a twenty-four-hour day, however, they have also been shown to work when daily cycles are as short as twenty-one hours or as long as twenty-seven hours. Normal twenty-four-hour cycles of light and dark have to be created in submarines, above the Arctic Circle and below the Antarctic Circle when the Sun is down or up for weeks or months, and in spacecraft.

The problem is that humans cannot function well if the cycle is shorter than about twenty-one hours or longer than about twenty-seven hours. In such environments our ability to do the things that we normally do each "day" breaks down. Behaviors change, bodily functions become erratic, and in general people go to pieces. So, don't go to Dichron.

EVOLUTION ON DICHRON

Plant Life

The spread of life over Dichron's lands will be profoundly different than occurred on Earth. Geologists have shown that because of plate tectonics, the continents on Earth have at different times all been united in one super-continent. The most recent incarnation of this massive land, called Pangaea, lasted from about 300 million years ago to about 180 million years ago, when it split into two smaller supercontinents: Gondwana, from which Africa, South America, Antarctica, Australia, Madagascar, and related lands evolved; and Laurasia, from which North America and the European–Asian land mass formed.

By the time Pangaea split apart, the evolution of land animals and plants on Earth was well under way. As a result, biologists find fossils of animal and plant life on widely separated continents today that originated from single species on Pangaea. The continents were essentially where they are today when homo sapiens sapiens (i.e., our "human" ancestors) evolved in east Africa about 200,000 years ago. These folks spread to different continents and changed (e.g., in skin color, facial features, and digestive capacities) in response to the environments in which they settled.

The major reason life evolved differently on Dichron than on Earth is that for as long as they exist, Dichron's continents will be stationary with respect to each other. Because they will never be in contact, it will be impossible for life-forms to walk directly from one continent to another, as occurred on Earth.

Consider a plausible scenario for the development and spread of life on Dichron: as on Earth it is likely, verging on certain, that life began in Dichron's oceans. Spread by ocean currents, plant life came to inhabit vast tracts of the oceans, including the shorelines of all three continents. This dispersion of ocean life makes it plausible that Dichron's three continents developed plant life at around the same time. Because of the different geography and climates these plants encountered on the three continents, the plants on the three continents of Dichron would evolve differently, creating three distinct families of life there.

As an alternative scenario, it is possible that plant life got a toehold on one continent millions of years before it moved onto the other two. We see islands on Earth thousands of miles from continents that have never been parts of larger land masses, but which had vegetation when first visited by humans. Seeds from established plant life on continents drifted on the oceans fantastic distances and then germinated on the islands. Plants that are known to have achieved this impressive feat of oceanic travel include coconuts, mangos, almonds, and mangroves, among dozens of other species.

I assert that both mechanisms for the establishment and spread of plant life on the continents of Dichron occurred. First, I expect that the separate, fundamentally different plant species evolved on each continent from ocean plant life in their vicinities. Then, over millions of years, sophisticated seed pods, such as coconuts, developed the capacity of surviving for years in the oceans and made their way to other continents.

Animal Life

Animal life could plausibly evolve separately on Dichron's three continents, but I am taking the liberty of restricting the initial establishment of the animal kingdom there to only one of them. The challenge is to get it to the other two land masses. Based on how life has spread to islands on Earth, there are at least two ways that would happen: first, floating seeds that spread over the oceans of Dichron could serve as rafts for small animal life. The animals could deposit larvae in the seeds (think, coconut), which float in the currents for months or years. During that time, the larvae develop and transform, eating the nutrients in the seed, and then boring their way out. Depending on the gestation time of the animals, some of the seeds would plausibly traverse the ocean between continents before all their nutrients are consumed and the pod sinks for one reason or another. Once on the shore of the new world, the animals could hop off, reproduce, flourish, and evolve there.

The second way life could spread between continents is via birds. As we know from birds on Earth such as godwits, which travel between Alaska and New Zealand, it is possible for some of our feathered friends to travel the distances necessary to go from one continent to another on Dichron. They can deliver themselves to the different continents, of course, but they can also carry other life with them in their intestines or in their feathers.

Once life is established on one of the continents, it will begin evolving to best adapt to its environment, a process that would continue for millions upon millions of years. Although other life will episodically drift in from the other continents, the life that adapts most effectively to the conditions that exist on each landmass is going to dominate. It is likely, therefore, that as more and more complex life develops on Dichron's continents, that life will diverge greatly from its ancestors. By the time sentient people begin exploring the different continents, they will discover profoundly different species on each.

If different types of sentient life evolve separately on each of the continents, it is unlikely that people from one continent would even recognize the sentient life-forms on another. They would all, however, have one thing in common: somehow they would all know when the land under their feet was about to melt.

5

What If the Earth Formed Fifteen Billion Years from Now?

FUTURA

A Brief History of the Rediscovery of Ground Zero

By M. T. Snerd, FRAS

The "War to End All Wars" on Futura was nearly the war to end all life. For centuries, the survivors eked out lives farming, fishing, and hunting. The edible animals were nearly eradicated before a semblance of civilization returned. It took 650 years after the war petered out for another modern, high-tech society to develop. It was during this heady period of rebirth that the young archaeologists Samuel Klein and Audrey Mills, and their students, explored what later became known as the Ground Zero complex.

Clearly a relic of times past, Ground Zero was a salt-covered plain about sixteen miles long and four miles wide. Tests revealed that it had once been the bottom of an ocean. It had fourteen disklike regions each about fifty yards across that had been melted at temperatures of over 12,000°F. It also had forty-one sets of parallel grooves nearly a yard deep and a yard wide gouged into it. Each set was comprised of three grooves. These striations all ran roughly the same distance of about eight miles.

Strewn in mounds of various heights were treasure troves of ancient technological artifacts. Amazingly, from the perspectives of the archaeologists, the ages of these objects ranged from the "War" back at least 4,000 years before it. Sam died before the excavation was a tenth of the way to completion, but as he had spent most of his time working to understand just a dozen of the most complete artifacts, his work was well advanced and eventually led to several crucial breakthroughs about the society that had developed them.

Audrey and her students pressed on, providing a rich theory of the site and the people behind it. She continued this work until she turned eighty.

The final paper Audrey and sixteen of her students co-authored was published in the *Journal of Archaeology*. In it, they proposed that Ground Zero had originally been a surface salt mine, which accounted for the parallel grooves. As their ancestors had become technologically advanced, they had experimented first with high explosives, accounting for the circular remnants, and then with manufacturing of electronics and related items, as indicated by the technological artifacts on the same sites. Absolutely every conclusion in this paper was wrong.

In the two centuries following this work, the people on Futura developed the technology to travel into space. This leap was greatly expedited by the work of Sam Klein. Naturally, the moon was their first destination. It was there that they discovered the fleet of spacecraft, parked in tethered synchronous orbits behind that world so that they were not visible from the surface of the planet. Great debate ensued as to what world their ancestors had planned to visit. The Futurans knew by then that their world was the only one orbiting their star that could support life. Clearly, these craft were intended for interstellar voyages.

This discovery inspired a cottage industry of finding inhabitable worlds orbiting other stars, with the thought in mind of carrying out the missions of their ancestors. At the same time, it created the field of astroarchaeology, sending eventually over a hundred of these professionals to the ships, trying to understand the technology, history, languages, and intended social structure that was going to be used to keep the space travelers functioning constructively on their journeys.

Clearly the levels of engineering and scientific knowledge and sophistication possessed by their ancestors were far beyond their own. As the layers of technology on the ships were revealed, it became clear that the people who built them understood deep issues related to interstellar space travel. These included, among many other things: the hazards of radiation in space; the need to recycle virtually everything out there, including dead bodies; the need to harvest even the rarified gases in space; the differences in planetary properties that exist between inhabitable worlds and the need to adjust to them; the needs to maintain physical health, including developing new medicines during the voyages, and mental health, including keeping people busy even though they were destined to live their entire lives in space (inasmuch as interstellar trips would take several lifetimes).

It was on one of these ships that part of the mystery of Ground Zero was solved. In one of its cargo holds were discovered two mammoth spacecraft with retractable wings and with wheels that were exactly as wide as the impressions on Ground Zero. Furthermore, the separation between the wheels exactly matched the separation between the three grooves in each set of striations found there. The conclusion was that these were shuttle craft. It quickly followed that the round indentations were launch pads, from which shuttles blasted off to the interstellar craft.

The final piece of the puzzle was solved during a three-day symposium held at Ground Zero to discuss these findings. It occurred when Noblie Prize Laureate Doctor Professor Leopold Vaklempt was presenting a paper before a thousand space scientists on how the interstellar spacecraft had been manufactured (in pieces on the planet and then assembled in space), when suddenly he stopped, looked up, looked down at his notes, and then started hyperventilating. Someone in the audience, a retired EMT, ran up with a paper lunch bag, into which he instructed Vaklempt to breathe. Several minutes passed before the professor could speak. "I'm sorry. I'm sorry," he began. But before he could go on, it happened all over again.

When he was finally able to talk, he said to the assembled researchers and media, "I think...I think we have been missing something all these years. I can't understand how...yes, I can. But it is so obvious to me now. Those ships up there are not intended to take people to other worlds. They brought people here." He stopped. Utter silence. "Don't you see? Our ancestors came from someplace else... some other planet in some other star system. They...we...are aliens on Futura."

The universe, which is defined for our purposes as all space, time, matter, and energy that will ever have any effect on Earth, "came into existence" about 13.7 billion years ago. Earth, along with the rest of our solar system, formed about 9 billion years later. That delay of 9 billion years is not a magic number. Our solar system could have formed billions of years earlier or later than it did. In this chapter, after exploring how we know the age of the universe and of the solar system and how the universe is changing, I consider the consequences of living on a planet, Futura, in a solar system formed 15 billion years in the future. Presenting such an age as plausible also requires that I justify the premise that the universe is going to last that long, which I do.

THE AGE OF THE UNIVERSE

Whenever you look at something, anything, you are looking back in time. This is so because the light coming to you travels at a finite speed of about 186,000 miles per second. You are reading this book as it existed a billionth of a second ago. You see the Moon as it was about a second ago, the Sun as it was eight minutes ago, Sirius, the brightest star in the night sky, as it was 8.6 years ago, and the most distant stars visible to the naked eyes as they were 3,000 to 4,000 years ago. All the individual stars and small groups of stars, such as the Pleiades and Hyades, that we see with our naked eyes are members of our Milky Way Galaxy. A galaxy is an ensemble of between about ten million and ten trillion stars, along with varying amounts of interstellar gas and dust, and other, often exotic, objects all held in orbit together by their mutual gravitational attraction.

In the southern hemisphere, two other galaxies, the Large and Small Magellanic Clouds at distances of 160,000 and 200,000 light-years* from

---

* A light-year is the distance that light travels in one year, which is roughly six trillion miles.

us, respectively, are visible to the naked eye. In the northern hemisphere, we can see with our naked eyes the Andromeda Galaxy (the individual stars in it are too dim to make out without a telescope). We are seeing it as it was two and a half million years ago.*

Telescopes enable us to see more distant objects, meaning that with telescopes we can look even farther back in time. Between our nearest-neighbor galaxies and those ten billion light-years away we see many galaxies with a remarkably limited variety of shapes and sizes. Over that distance the shapes and sizes of galaxies don't change, meaning that the ones ten billion light-years away look similar in size, shape, and number to the ones that are very close to us (only millions of light-years away). Because we see the farther ones as they were farther in the past, we (correctly) conclude that most galaxies persist relatively unchanged for billions of years.

Carrying this argument further, if the universe has existed forever, we would expect to see similar galaxies even farther away than ten billion light-years. But beyond that distance, or equivalently looking back farther than ten billion years into the life of the universe, we see galaxies that are fundamentally different than the closer galaxies. For example, galaxies thirteen billion light-years away (meaning that we are seeing them as they were some thirteen billion years ago) are not nearly as well developed as those that are closer to us. We see at such distances and beyond, among other things, clumps of stars and intergalactic gas in the process of merging to create full-fledged galaxies. This suggests that the universe was fundamentally different thirteen billion years ago than it is today. Just how different becomes clear momentarily.

Another clue about the history and age of the universe came on the heels of Albert Einstein's 1915 publication of the theory of general relativity. This work boils down to a set of ten equations that describe the behavior of matter under the influence of gravity and the other forces in nature. The results of calculations done using general relativity are more accurate than the same calculations done using Newton's three force equations. The cost is considerably more numerical manipulation.

In 1922 Russian mathematician Alexander Alexandrovich Friedman

---

* The Triangulum Galaxy, 2.8 million light-years away, can also be seen with the naked eye, but only barely and only under ideal conditions.

found a solution to Einstein's equations that predicted the expansion of the universe. This was followed in 1927 by further calculations done by Georges Henri Joseph Edouard Lemaître, Belgian priest and physics professor. In 1929, American astronomer Edwin Powell Hubble was the first person to observe that all but the nearest galaxies, such as Andromeda and the Magellanic Clouds, are moving away from us. His observations have been confirmed by numerous astronomers since then: the universe is expanding. The expansion of the universe was discovered because objects change color as they move apart, an effect called the Doppler shift.

Doppler Shift

Light is an intriguing phenomenon because it travels as particles called photons that have intrinsic wave properties (Figure 5.1). When we think of particles, pieces of sand or dust come to mind. Examining them under a microscope, one sees solid objects. Conversely, when we think of waves, water waves are an obvious example. Once they pass, the water settles back down without leaving a trace of the wave. Photons have properties of both particles and waves. Just as you feel the impact when you catch a baseball, a sufficiently sensitive sensor feels the impact of a photon. At the same time, the wave in each photon has a single wavelength associated with it.

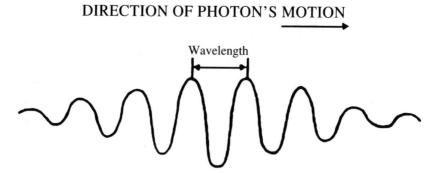

DIRECTION OF PHOTON'S MOTION

Wavelength

Figure 5.1: Conceptual drawing of a photon, showing both wave and particle properties. The distances between consecutive peaks in a photon are all equal. This distance is the photon's wavelength.

Depending on its wavelength, we assign photons different names. From the longest wavelength photons to the shortest, we call them: radio wave, microwaves, infrared radiation, visible light, ultraviolet, x-rays, and gamma rays. These are listed in the same order presented in Chapter 2 for photons of increasing energy: the shorter the wavelength, the higher their energy. Different colors are created by visible light photons of different wavelengths. Red has the longest wavelengths,* followed by orange, yellow, green, blue, and violet with the shortest ones.

We see most objects because the particles in them scatter visible light of only certain colors. Red objects scatter red light, turquoise objects scatter blue and green light, and so on. The reason different objects have different colors is because they are composed of different atoms and molecules, and every different element and molecule emits a unique set of wavelengths. When a light source of, say, hydrogen is sitting at rest in the laboratory, if you were to spread out the colors it emits with a prism, you would see that the hydrogen actually emits only a few visible wavelengths (i.e., colors), notably one red, one green, and two shades of blue. It also emits photons at certain nonvisible wavelengths. No other element or molecule emits these and just these wavelengths. What follows applies to all types of photons, not just visible light.

If that same hydrogen gas were traveling away from you, all its colors would be shifted toward longer wavelengths. We call such a change a red-shift because the colors are shifted toward the longest-wavelength visible color, which is red. Conversely, if the hydrogen were traveling toward you, all its colors would be shifted toward shorter wavelengths. We call this blue-shift. (No, I don't know why it isn't called violet—equivalently, purple—shift, inasmuch as violet is the shortest wavelength of visible light. Perhaps it is because blue is easier to say.)

This change in wavelengths with motion toward or away from us is called the Doppler shift, after Austrian physicist Christian Andreas Doppler, who calculated it in 1842. The Doppler shift is analogous to the change in pitch you hear when a siren passes you. Suppose you are driving down the

---

* The "s" on "wavelengths" here is included because each fundamental color has different hues that correspond to different wavelengths. The wavelengths associated with the hues of each color are all adjacent to each other.

interstate at, oh, 75 mph when suddenly you hear a siren behind you. My theory is that if you don't look in the rearview mirror, the cop can't see you, so they can't be after you. As you slowly decrease to the legal speed and begin pulling over, praying that the cop is after someone else, the ambulance rushes past. Your adrenalin level drops dramatically. If you listen to the siren as it goes by, you will hear it change pitch to a deeper sound, a Doppler shift. This is analogous to a red-shift of light. Conversely, if an ambulance with its siren on is at rest relative to you and suddenly begins moving toward you, the siren changes to a higher pitch. This is also a Doppler shift, analogous to a blue-shift.

By studying the details of the light emitted by stars in distant galaxies, Hubble discovered that except for the nearest galaxies, all galaxies emit red-shifted light as seen from Earth. This means that they are all moving away from us, as stated above.

Measurements by Hubble and others also revealed that the speeds of those galaxies that are moving away from us depend on how far they are. Galaxies twice as far are moving away twice as fast, those ten times farther are moving away ten times faster, and so on. Astronomers also discovered that the speed away from us at a given distance is the same in all directions.

So, all distant galaxies are moving away from us symmetrically. When I first heard that, it seemed to imply that we must therefore be at the center of the universe. It doesn't and we aren't. This brings us to the modern astronomical understanding of the history and evolution of the universe, which explains its expansion and how it is that we aren't in the center even if nearly everything is moving away from us.

A scientific theory that describes the evolution of the universe is called a cosmology, and the one (and only one) such theory that accurately predicts everything that we observe is called the Big Bang.* The Big Bang cosmology

---

* Please don't get the impression that this theory is complete. It is not. However, thus far all observations we have made of the universe are consistent with the equations that describe this cosmology. Furthermore, when observations are made that the theory does not predict in its present form, the equations can be modified without becoming inconsistent with each other. For example, observations recently revealed that the universe is expanding faster and faster, which the Big Bang in its previous form did not predict. However, changing its equations to include outward acceleration maintained a consistent cosmological theory. Therefore, the Big Bang theory is continually being refined and made more consistent with the observed nature of Nature.

is a scientific theory* that is consistent with all observations of the structure and evolution of the universe that have been made thus far.

Big Bang Cosmology

In the beginning . . . let me begin by acknowledging that astronomers do not yet know how the universe came into existence some 13.7 billion years ago. However, the Big Bang theory provides equations and descriptions consistent with the proposed and observed evolution from about one-tenth of a millionth of a billionth of a billionth of a billionth of a billionth of a second (more precisely, 0.00000000000000000000000000000000000000000054 sec = 5.4×10^{-44} sec) after it began through today and into the future. For future reference, 5.4×10^{-44} seconds is called the Planck time, after German physicist Karl Ernst Ludwig Max Planck (Max to his friends, 1858–1947), who first calculated it.

Here is an outline of the Big Bang cosmology that will provide the information we need to justify the premise of this chapter that an Earthlike world could be formed fifteen billion years from now, and that it would have a different chemistry than our planet.

Before the universe began there was no space, no time, no matter, and no energy as we know these concepts. The universe began as an explosion that created all of these things. Expanding at a rapid rate, matter and energy were initially combined into one unified lot. Back then there were no such things as atoms, photons, gravity, electromagnetism, or other characteristics of matter or of the forces through which they interact. (Today there are four known forces in nature: electromagnetism, gravitation, and the weak and strong nuclear forces. We experience the first as the photons discussed earlier in this chapter. The latter two forces exist only inside the nuclei of atoms. They serve to mediate the interactions between the particles that exist in nuclei.)

A tiny fraction of a second after the beginning of time, the force we know of as gravity is believed to have separated from the other three, followed by the strong force, and then the electromagnetic and weak forces separated from

---

* A theory is considered scientific if it is based on equations that make predictions that can be tested, and the theory can potentially be disproved. A scientific theory differs from other theories that must be taken on faith, such as religious beliefs. These latter theories cannot be disproved.

each other. The formation of the particles that make up normal matter (protons, neutrons, and electrons) followed shortly thereafter. In the first three minutes that the universe existed, some of the newly formed protons and neutrons smashed together and stuck, a process called nuclear fusion, thereby creating some of the helium and about ten percent of the lithium that exist in the universe today. This early fusion occurred because the universe was incredibly hot when it first formed: at a time of 100 seconds, the temperature of space was a toasty 1 billion K (1.8 billion°F), compared to about 295 K (72°F) in a typical room.

The high temperature was what gave the particles enough energy to fuse together. As the universe expanded, it cooled. This is analogous to the effect you get if you heat an oven, turn it off, and then open it up. As the hot gases in the oven expand into the kitchen, the overall temperature of the gas decreases. After about three minutes of expanding and cooling, the temperature in the universe dropped to about 10 million K (18 million°F). Below this temperature, particles cannot fuse together, so the initial fusion of protons, neutrons, and electrons came to a halt. At that point, the universe was composed of roughly seventy-five percent hydrogen, twenty-five percent helium, and a trace of lithium.

Atomic Physics—A Primer

Each atom is composed of a nucleus comprised of protons and neutrons, often surrounded by a swarm of electrons. Protons have a property called electric charge. They are arbitrarily assigned a positive charge. Electrons have exactly the opposite electric charge, called negative. Particles with like charges (pairs of protons or pairs of electrons) repel each other, whereas a proton and an electron (i.e., opposite charges) attract each other. Neutrons have no charge. They provide some of the "glue" in atomic nuclei that keeps the protons, which are continually repelling each other, bound together.

The type of *element*, such as hydrogen, helium, lithium, carbon, oxygen, iron, and the like, is uniquely determined by the number of protons in an atom's nucleus. Every atom in the universe with one proton is a hydrogen atom, every atom with two protons is helium, every atom with three protons is lithium, and so forth.

Many elements can have different numbers of neutrons. For example, hydrogen can have no neutrons (just a single proton), one neutron bound to the proton, or two neutrons bound to the proton. Elements with different numbers of neutrons are called different isotopes of the element.

Electrons can orbit nuclei only in certain "allowed" orbits that depend on the type of element and to which isotope the electron is bound. If the number of electrons orbiting around the nucleus equals the number of protons in that nucleus, then the atom is electrically neutral. Otherwise, it is electrically charged and is called an ion.

By absorbing photons, electrons gain energy and can thereby be made to change orbits or even to be yanked out of orbit completely, a process called *ionization*. Some photons have the right amount of energy to transfer electrons from one allowed orbit to another, higher-energy, allowed orbit. These photons are absorbed by the electrons, becoming part of them. Photons that do not have the right amount of energy to transfer electrons from one allowed orbit to another or to ionize them zip right through the atom.

Electrons in higher-energy orbits are often unstable, cascading down to lower-energy orbits. In doing so, they emit photons with energies equal to the difference between the energies of the orbit the electron left and the energy of the orbit to which it fell. As noted earlier in this chapter, the energy of each photon is determined by its wavelength. Every different isotope of every element has a unique set of allowed energy orbits. Therefore, each transition in each type of isotope leads to the emission of a photon with a unique predictable energy and wavelength. These are the fingerprints of all the isotopes of all the elements in nature. They explain the colors in the discussion of Doppler shift, above.

How We Are Not in the Center of the Universe

In addition to the matter and energy we have been discussing, the Big Bang explosion also created space and time. Einstein, in developing his special* and general theories of relativity discovered that space and time must be

---

* Whereas general relativity allows for the presence and effects of gravitation, special relativity deals with gravitation-free, constant-speed motion. It is useful when the force of gravity is negligible compared to other effects.

treated on equal footings. In other words, whereas we normally think of space having three dimensions (front and back, up and down, side to side), in physics, time becomes a dimension and we must deal with spacetime as four dimensions.

Spacetime immediately started expanding after the Big Bang. The crucial concept here is that every point in spacetime was and is moving away from every other point. You can visualize this by imagining a spherical sponge. Imagine three equally spaced dots placed in a straight line inside the sponge. Now crush the sponge tightly into a small ball. The sponge then represents the three dimensions of space when the universe was young. All places in the universe were initially very close to each other. To represent time moving forward (and the universe growing) allow the sponge to uniformly grow. Every point in the sponge moves away from every other point in it.

Imagine yourself on one of the outer of the three dots inside it. This dot represents our local neighborhood of the universe and as the sponge expands, every point in it moves away from that dot. In order for the sponge to expand uniformly, places farther away from the dot in the sponge must move away from it faster than do places closer to it. For example, in order for the farther of the other two dots to always remain twice as far away from you as the closer one, the farther one must move away from you twice as fast as does the closer one. This kind of motion is just what Hubble observed for the expanding universe: galaxies twice as far away are moving twice as fast away from us. Now put yourself on either of the other dots and visualize the same expansion from there. You get the same results. Every place in the universe sees the rest of the universe expanding away from it, just as we do here on Earth!

Explaining the Doppler Shift

The Big Bang cosmology neatly explains the Doppler shift that we see for distant galaxies. Because spacetime (the universe) is expanding, it carries with its expansion all the distant galaxies. In other words, we see them moving away from us as markers of the universe's expansion.

It is worth mentioning why I keep saying "distant galaxies." Individual galaxies, held together by the gravitational attraction of all the things in them, are not expanding. Likewise, as the universe expands, galaxies that are near each other are so strongly bound by their mutual gravitational attraction that this expansion is not strong enough to pull them apart. These gravitationally bound galaxies orbit each other as the universe expands around them, a fact that comes back to haunt us shortly.

Why We See Galaxies Appear Younger at Great Distances from Here

The next set of observations that the Big Bang cosmology must explain is why, at distances of roughly 13 billion light-years and more, galaxies appear less well-formed than those seen at closer distances. Using the results of Hubble and numerous astronomers since his pioneering work, we have an equation that relates the distance to distant galaxies and the speeds at which they are moving away from us. Ask the question: suppose that we reverse the flow of time so that the universe is contracting, rather than expanding. How long would it take until we return to the extremely dense, hot state of the Big Bang? The equation relating expansion speed and distance run in reverse tells us the answer, which is about 13.7 billion years. This is how we got the age 13.7 billion years for the universe as introduced earlier.

During the first 700 million years of the universe's existence, the hydrogen, helium, and traces of lithium gases created in the Big Bang clumped together and formed stars and bound groups of stars that merged and became galaxies. By looking back that far in time (i.e., that far away from us), astronomers are now becoming able to watch these events, which is what we see beyond 13 billion light-years away. Keep in mind that observing at such great distances is pushing the limits of our technology. We have barely scratched the surface of knowledge about that realm of the universe's very early history.

Cool It

To help show the power of a theory like the Big Bang, let me consider the consequences of the high temperature that the theory predicts for the young universe. The expanding universe is cooling, as the heat generated in the Big Bang explosion is spreading out. Such an observation led astrophysicists Robert Henry Dicke and Philip James Edwin Peebles at Princeton

University to propose in the early 1960s that the universe should not be completely cold; there should be remnants of the radiation that made the universe 1 billion K when it was 100 seconds old. Dicke and collaborators began looking for those remnants, not knowing that two scientists at nearby Bell Laboratories had already discovered them.

Robert Woodrow Wilson and Arnault Allan Penzias were trying to develop a high-sensitivity receiver so that Bell Labs could transmit telephone signals* without irritating noise that was common in radio telecommunication back in those days. Wilson and Penzias could not get rid of all the noise they heard. The reason was that they were hearing radio signals from space that were remnants of the superhot radiation created at the beginning of time.

The electromagnetic radiation that heated the early universe was originally composed of extremely short wavelength gamma rays, the most powerful photons that can exist. As the universe expanded, the wavelengths of each remnant photon expanded with it until today, according to the calculations of Dicke and Peebles and consistent with the observations of Wilson and Penzias, most of those photons should be microwaves. One can go further and calculate the temperature that the universe should have today as a result of its expansion and consequent cooling. The calculated temperature was 2.735 K. The temperature was measured by the COBE (COsmic Background Explorer) satellite in 1990. It was 2.735 K.

Will the Universe Last for Thirty Billion Years?

I have established that the universe began. We can watch galaxies at different distances from us, so that we have a pretty good understanding of how it is evolving. Before considering that evolution and how it relates to our later version of Earth, Futura, I want to address the question of whether the universe is going to last long enough for us to consider the scientific consequences of an inhabitable world forming fifteen billion years from now.

The fate of the universe hinges on how much matter and energy it contains. Within a minute of when the Big Bang explosion occurred, the gravitational attraction of all the matter and energy in it began slowing down its

---

* This was WAY before cell phones were even a twinkle in some engineer's eye.

expansion. This is the same effect you experience when you jump upward. The gravitational attraction of the Earth pulls you back down. If you could jump up fast enough (about 25,000 mph) you would travel upward into space and not fall back. This is called the Earth's escape velocity.

If the universe has too little mass to provide enough gravitational attraction to stop the expansion, then its components have escape velocity with respect to each other, assuring that the universe will expand forever and justifying our exploration of the future Futura. If, on the other hand, the universe has too much mass, we would expect, all other things being equal, that the gravitational attraction of the matter and energy will eventually stop the expansion and cause it to collapse. Calculations indicate that even if the universe were going to collapse, it is still moving outward fast enough so that the stopping and collapsing would not occur for at least another 30 billion years. A calculated lifetime of 120 billion years is typical for the universe if it were going to collapse. That is encouraging for the existence of Futura. But there is another factor that makes it a dead certainty that we are justified in studying Futura. It was discovered in 1998.

With this issue of the fate of the universe in mind, many astronomers spent much of the twentieth century trying to identify all the matter and energy in it to see whether there was enough gravitational force to stop the universe's expansion. Matters took a totally unexpected turn in 1998, when they discovered that the universe is actually accelerating outward. Not only is it never going to stop expanding, it is expanding faster and faster and, as far as we can tell, will continue to do so.

At present, we do not know the cause of that outward acceleration because it requires a force that pushes against the attraction of gravity. The three other forces (electromagnetism, and the weak and strong nuclear forces) do not have repulsive (i.e., outward-pushing) effects that can overcome gravitational attraction at such large size scales.* Einstein's equations that describe the Big Bang cosmology do, however, allow for the existence of a variety of other repulsive forces. Astronomers are now trying to determine which of these forces exists and causes the universe to be accelerating.

---

* Although the nuclear forces act only at size scales of 10^{-13} inches, the effects of electromagnetism do extend over the size of the entire universe. However, because there is one positive electrical charge for each negative charge, the electromagnetic effects cancel out over the scale of the universe.

Armed with the knowledge that the universe is going to continue getting larger, we now have the scientific justification for examining how it will change over that time and, therefore, how Futura will be different from the Earth. First, we need to decide where in the universe a later version of Earth, Futura, can form.

Evolution of Stars in Spiral Galaxies

As we saw in earlier chapters, stars form from interstellar clouds of gas and dust. When small fragments of those clouds become suitably dense, their gravitational attraction causes the fragment to collapse. One or two stars typically form in the center of this collapsing matter. If the collapsing gas and dust are swirling, as most are, then they form a disk of debris around the stars in which planets form.

Most star formation in the universe occurs in galaxies with spiral arms. There are basically five types of galaxies, differentiated by their shapes (Figure 5.2). Two disk-shaped kinds of galaxies have spiral arms. One of these spiral galaxy types, called barred spiral galaxies, has a bar of stars, gas, and dust running through its middle. Our Milky Way is a prime example of a barred spiral galaxy.

The other type of spiral galaxy, those without a bar, is called a normal or unbarred spiral. As I discuss in more detail shortly, both types of spiral galaxies form stars in their spiral arms. Many of these stars eventually explode, returning gas and dust into the interstellar medium. This debris is then used to form new generations of stars, a cycle that can repeat, with some modification, for tens of billions of years. The third type of galaxy is disk-shaped, but without spiral arms. These are called lenticular.

The fourth type of galaxy, the ellipticals, appear to us as ovals of stars. They are actually spherical or egg-shaped, and they undergo much less star formation than do spirals because elliptical galaxies contain much less interstellar gas and dust from which to create new stars. They have less of this material because after the first generation of stars forms, the gas and dust shed by exploding stars doesn't remain concentrated in the galaxy, as it does for both kinds of spiral galaxies. The fifth type of galaxy, called irregular, has no consistent global shape. They appear, well, irregular. These also have much less star formation occurring than do spirals. Given the options, I

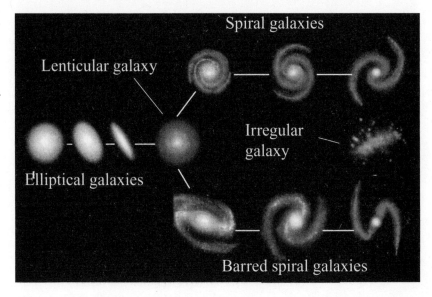

Figure 5.2: *Five different shapes of galaxies in the cosmos.* CREDIT: W.H. FREEMAN & CO.

choose to have Futura form in a barred spiral galaxy, but for reasons that become clear in Chapter 10, it won't be our Milky Way.

EVOLUTION OF FUTURA'S GALAXY

I call the galaxy in which Futura forms the Musketeer. It is a barred spiral galaxy like the Milky Way containing about 500 billion stars accompanied by interstellar gas and dust amounting to about fifteen percent as much mass as is in the stars. To understand how Futura will differ from Earth, let's begin by following the evolution of the Musketeer.

Why Spiral Galaxies Have Spiral Arms

When a spiral galaxy first forms, it does not have spiral arms. Visually, it is a disk of gas, dust, and stars: lenticular. A variety of things can create spiral arms. For the Musketeer, as for the Milky Way, it was the passage of a much smaller galaxy. In our case, there are several such galaxies, including the Magellanic Clouds. Imagine throwing a rock in a calm lake. The ripples, of course, will be concentric rings spreading out from the impact site. If the lake were spinning . . . that's hard to envision, so consider a pie pan filled with water sitting on an old-fashioned turntable. Turn the turntable on at 78 rpm (revolutions per minute) and then, when the water has stopped sloshing around, drop a pebble into the water. Instead of having concentric rings of water spreading out from the impact site, the spinning water creates spiral ripples, with the arms wrapped so that the tips point opposite to their direction of motion. This is the basis of the spiral arms in many spiral galaxies.

In spiral galaxies, the spiral arms are sound waves of slightly higher density (exactly like sound waves in the air) that travel through the gas and dust that orbit around in the disk. These so-called spiral density waves travel about half as fast as the gas and dust through which they move. Therefore, the gas and dust from behind the spiral arms pass through them. As this gas and dust go through a spiral arm, this debris is compressed by the pressure from the wave, causing some of the gas and dust to collapse and form stars. Indeed, the vast majority of all the stars created in most spiral galaxies are created this way.

The stars that are formed have a variety of masses, from 0.08 M⊙ (0.08 solar masses, meaning $^{8}/_{100}$ times the mass of the Sun) up to around 100 M⊙ As we show momentarily, stars with different masses evolve differently and explode at different times. The more massive a star is, the more brightly it shines, the faster it evolves, and the more powerfully it explodes. The reason we see the spiral arms, which are after all just regions of slightly denser (like five percent denser) gas and dust, is because the brightest stars formed in them are so bright that they cause the surrounding gas to glow brightly, highlighting the arms. These brightest stars, living only a few million years, explode before leaving the arms. The stars that survive their passage through the arms (remember that the stars were created from gas moving twice as fast as the arms themselves) are so dim that they don't make the gas and dust between arms glow very brightly.

Why Stars with Different Amounts of Matter Have Different Lifetimes

As just noted, more massive stars live for much shorter lengths of time than do less massive ones. A 25 M⊙ star will shine for about eight million years, whereas our (1 M⊙) Sun will shine for about ten billion years. It is important to understand the evolution of stars and how they change the chemistry of the interstellar medium, the gas and dust from which future generations of stars, including Futura's star, are formed.

Recall from earlier in this chapter that the Big Bang explosion caused the early formation of hydrogen, helium, and lithium. However, there are 92 naturally occurring elements today. Therefore, all of the atoms of the other 89 elements, as well as some helium and most of the lithium that exist today, had to have been created after the Big Bang. In fact, they were, and still are, created deep inside stars. Here is how that works.

When a small piece of an interstellar cloud of gas and dust collapses to form a star, that matter is mostly hydrogen and helium that had been created in the Big Bang. As this gas piles up on itself to form the star, the gravitational attraction of all this matter compresses the center of the star. This compression causes the temperature of the gas in the core to soar. When it exceeds 10 million K (18 million°F), the hydrogen atoms can fuse together, creating helium, just as was done during the Big Bang.

This fusion in the star's core is accompanied by the release of energy in the form of gamma ray photons.* Colliding with other atoms, these photons lose energy until they emerge from the star's surface layer as starlight. The energy that we see as sunlight, for example, was created by fusion in the Sun's core over 100,000 years ago.

So, the fusion occurring in stars like the Sun is converting core hydrogen into helium. When the core is pure helium, it begins fusing into carbon and oxygen. At the same time this is occurring, a shell of hydrogen surrounding the core will be fusing into helium, and that helium will eventually fuse into carbon and oxygen, too. After the core is all carbon and oxygen, stars with masses similar to that of the Sun shed their outer layers, now enriched with helium, carbon, and oxygen from the fusing shells, in mild "explosions" called planetary nebulae. Clearly, the fusion that occurred in the star results in the debris that returns to the interstellar medium having less hydrogen than the gas from which the star formed. The cores of such stars remain intact. They are called white dwarfs.

---

* Neither mass nor energy is conserved in this process. Specifically, mass is converted into energy in the amount: (Energy created)$=$(mass lost)$\times c^2$, where c is the speed of light. The sum of (mass$\times c^2$) plus energy is conserved.

Elements with more protons, called heavy elements, are created in stars with more mass than about 9 M⊙ As with lower-mass, Sunlike stars, fusion begins in their cores, converting hydrogen into helium. When the core hydrogen is used up, a shell of hydrogen surrounding the core begins fusing hydrogen into helium, and the core fuses its helium into carbon and oxygen. When the core helium is used up, its carbon and oxygen fuse into silicon. At the same time, the helium created in the shell outside the core begins fusing into carbon and oxygen, and another shell of hydrogen outside this shell begins fusing into helium. This process of fusing into heavier and heavier elements continues until the inside of the star resembles an onion, with the heaviest element being created in the core and successively lighter elements being created in shells outside it (Figure 5.3).

Eventually the core is transformed into iron, with twenty-six protons and thirty neutrons in its nucleus. This core is about the size of the Earth. Iron cannot fuse into heavier elements, so a dramatic change in the evolution of the star occurs. The iron, compressed by the matter above it, condenses until there is enough pressure and temperature in it to force the electrons that were outside the iron nuclei into them. These electrons fuse with protons to form

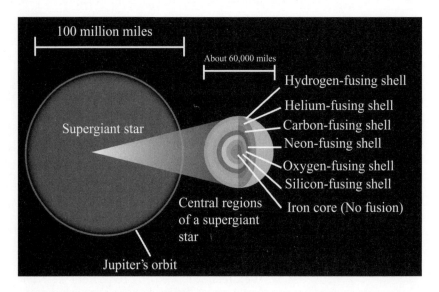

Figure 5.3: Shells of fusion in the central region of a supergiant star. CREDIT: W.H. FREEMAN & CO.

neutrons. Without the protons defining iron, the core is quickly converted entirely into neutrons.

Under the pressure it experiences, a neutron core collapses at roughly a quarter the speed of light. In less than a second it has collapsed from the size of the Earth to a sphere only sixty miles across! At that point, it is so dense that the neutrons smash into each other and stop the collapse so vigorously that the core literally bounces back outward at nearly the same speed. This bounce is similar to the bounce you see when dropping a bouncy ball onto a hard floor.

During the second when the core is collapsing and bouncing, the outer layers of the star also begin falling inward, as they are no longer supported by the core. The outward-bouncing core and outward-flowing particles called neutrinos created in the core meet the inward-falling matter. The resulting collision causes the outer layers, now rich in many elements from the shells of fusion, to begin rushing outward in a titanic explosion called a supernova. At peak brightness a supernova emits ten billion times as much light and other energy as the Sun. Supernovae get nearly as bright as all the stars shining in their galaxy, combined!

The elements created by fusion during the life of the star are ejected in the supernova, along with other elements created during the supernova explosion itself. These latter include many of the naturally forming elements from iron through uranium.

For completeness, it is worth mentioning that there is another major mechanism for supernovae. This involves carbon-oxygen white dwarfs at the upper limit of the mass that these remnants can have. If more gas falls on these maximum-mass white dwarfs, say from a companion star, they collapse in on themselves, start fusing wildly, and explode in even more powerful explosions than the supernovae just discussed. The supernovae from white dwarfs generate and send into space a lot of iron and related elements. Although occurring less frequently than the type previously discussed, supernovae from white dwarfs do contribute measurably to the elements in interstellar space.

There are still some naturally forming elements that are neither created inside stars nor in supernovae. These remaining elements come from the radioactive decay of some elements ejected by supernovae. Radioactive elements are unstable, spontaneously splitting into two or more pieces. For example, uranium is fused into existence in supernovae. It is radioactive, so it spontaneously decays, as do some of the elements it forms. This process eventually creates thorium, radium, radon, lead, and bismuth, among other elements. In summary, all the elements found in nature are created as a result of stellar fusion and explosions, and by radioactive decay.

The cores of stars that remain after planetary nebulae and supernovae are remnants of their glory days. As noted earlier, the remnant of the Sun, a carbon-oxygen body that is no longer fusing anything, will be a white dwarf. The remnants of stars that initially had between 9 and 25 M☉ are neutron stars.* The remnants of stars that initially had more than 25 M☉ will end up as black holes, created because the gravitational attraction of the neutron core on itself is so great that it overcomes all repulsive forces and causes the neutrons to collapse until they are in such a tiny volume of space that no matter or energy, except gravity, can escape from it: a black hole.

The fraction of a high-mass star's mass that is injected back into interstellar space can be as much as ninety-five percent of its original mass, although for lower-mass stars the amount returned to circulation is much less. For example, the Sun, an average-mass star, will only shed about sixty percent of its outer layers.

Reprocessing Stellar Remnants into New Stars

Over the years, stars form, evolve, and finally shed their outer layers. The matter that they send out is a combination of gas and dust, which is now richer in heavy elements (called metals†) and poorer in hydrogen and helium than was the gas from which the star originally formed. These metals, orbiting in the galaxy with the remaining hydrogen and helium, are used in the next generation of stars and planets.

One of the unsolved mysteries of galaxy evolution is how much of the gas and dust from explosions remains in the disk of spiral galaxies and how

---

* Bad name, as the neutrons are no longer fusing or otherwise creating new photons. All the light they emit comes just from the heat stored in them. As they cool, neutron stars dim.
† Although highly nonintuitive, all elements in nature except hydrogen and helium are classified as metals by astronomers.

much fresh gas, mostly hydrogen and helium, falls onto the disk from outside it. The more gas that falls in, the more dilute in metals becomes the gas that forms new stars. Because I want to explore a world with more heavy elements than ours, I posit that the Musketeer Galaxy has a considerable amount of metal-rich gas and dust from stellar explosions combined with a relatively low amount of hydrogen and helium gas falling onto its disk from its outer regions.

After gas and dust from supernovae and planetary nebulae are inserted into the disk of the Musketeer, that debris passes into a spiral arm, where some of it is compressed and forms the next generation of stars. Remember that shortly after the Big Bang occurred the universe was composed of 75 percent hydrogen, 25 percent helium, and a trace of lithium. Converting that matter into metal takes time. Lots of time. Today, 13.7 billion years after the universe began, the fractions of the most common elements are 74 percent hydrogen, 23 percent helium, 1 percent oxygen, .5 percent carbon, .13 percent neon, .11 percent iron, .10 percent nitrogen, and less than .1 percent of each of the other elements.

After 30 billion years, hydrogen and helium will still comprise the bulk of each star formed, but the disks of gas and dust surrounding stars from which planets and debris form (see Chapter 1) will be much richer in heavier metals and these elements determine the differences between that world and Earth.

FUTURA

As with Earth, the planet-building process that created Futura was based on countless collisions of gas and dust particles, which combined to create larger pieces of debris. Several pieces of this rubble, orbiting their newly formed star at different distances, became so large that the other debris they encountered was pulled onto them until entire regions of the space around that star were swept clean. One of these growing masses became Futura, and the others became other planets.

Futura is a world with the same mass and the same initial rotation rate as the Earth, orbiting a Sunlike star. However, Futura formed from a disk of gas and dust that was much richer in heavy elements than was the material from which the Earth was formed. Most notable among the dense heavy

elements are iron and nickel, along with greater amounts of chromium, tita-
nium, cobalt, copper, lead, gold, and platinum compared to what exist on
Earth.

Length of the Day

As it formed, Futura was a more or less uniform blend of different elements. But due to the heat generated by the collisions during that period and
from short-lived radioactive elements in it, the planet became molten. As it
grew, the denser iron and nickel in it sank toward the center, and lighter elements, such as silicon, oxygen, aluminum, calcium, sodium, potassium, and
magnesium rose toward the surface.

This process of differentiation (see Chapter 4) led to an early difference
between Futura and Earth. Recall from Chapter 3 that angular momentum
of isolated objects is conserved (think of the ice-skater pulling in her arms
and leg). We are considering Futura before the collision that creates its
moon, so the planet is basically isolated. Futura contains a higher fraction of
high-density iron and nickel than does Earth. In other words, Futura has
more matter that will sink in it than did Earth. Conservation of angular
momentum dictates that when the larger concentration of dense elements
sinks, the planet's rotation rate increases more than did the rotation rate of
the young Earth, with its lower fraction of iron and nickel. Therefore, by the
time the collision occurs that creates Futura's moon, the planet is spinning
faster than Earth was at the same stage in its evolution.

I assume that, as with Earth, a Mars-sized intruder hit the young Futura
and formed a moon within a few hundred million years after Futura formed.
After this collision occurred, Futura was still spinning faster than the Earth
at that time. Futura would slow down due to tides as the Earth did, but starting from a faster spin, Futura "today" would have a shorter day than we do. I
set Futura's day at the equivalent era as we are in today to be eighteen hours.

Futura's Interior

As with the Earth, Futura's core is divided into two parts: a solid inner
core and liquid outer core, both of which are composed primarily of iron.
Earth's inner core is about three-fourths the size of our Moon and has a

temperature that exceeds the Sun's surface temperature of 5,850 K (10,000°F). Because Futura has the same mass as the Earth and more heavy elements, both parts of the new planet's core will be larger and more massive than occur in the Earth.

Heat is generated inside Futura and other Earthlike planets in several ways. First is the heat created there when impacts compress the planets' interiors. Second is the heat created as planets differentiate. As the denser elements settle toward the core, friction generated by their motion creates heat. Third is the heat generated as the inner core of a planet grows. The heat released as metal on the surface of the core solidifies is called the latent heat of crystallization. Fourth is heat generated as the planet is compressed by its own mass. These all pale in comparison to the heat generated by the fifth mechanism, which is the decay of radioactive elements (see Chapter 4) in the planet's mantle.

The most common radioactive isotopes inside Earth and Futura are potassium 40, thorium 232, and uranium 238 and 235. These are all found in the mantle, but thorium and other radioactive isotopes may have also bonded to descending iron when the planets differentiated, causing them to be trapped in the planets' cores. Because it was created from materials that contained more radioactive elements than was Earth, Futura has more radioactive atoms in it than does our planet. The decay of these atoms provides most of its internal heat, thus Futura will have more convection in its mantle (see Chapter 4), delivering more heat to the crust, and therefore, Futura will have a more active surface than does Earth. I explore this further shortly. First, let's consider another important difference from Earth occurring deep inside Futura, namely the magnetic field that it generates.

Futura's Magnetic Field

Because Futura's (and Earth's) outer core is a liquid metal, heated by the hotter solid core below it, the molten metal convects, meaning that blobs of especially hot metal rise upward in the outer core, reach the bottom of the mantle, give their extra heat to the mantle rock, and sink back down. The rising and sinking metals have electrons that stream upward and then downward creating electric currents similar to those created by batteries to run electronic equipment, among other things. In 1819, Danish physicist Hans

Christian Orsted discovered that electric currents cause the deflection of compass needles. He correctly proposed that as it flowed, the current was generating a magnetic field, which in turn caused the deflection. The currents flowing in Futura's (and Earth's) outer crust also create magnetic fields. When convection in the core is combined with the planet's rotation, the magnetic field becomes a dipole (Figure 5.4), like a bar magnet, with one end of the field coming out at what we call the South Magnetic Pole looping around the planet in all directions and entering at the North Magnetic Pole.

Two major factors determine the strength of a planet's magnetic field: the amount of molten metal in motion in the core and the speed at which the planet spins (see Chapter 2). Futura's core has more molten metal and the planet spins faster than the Earth. Hence, Futura's magnetic field will be stronger than that of the Earth. Especially due to the extra internal heat, and hence the greater amount of convecting molten metal, I estimate that Futura's field is roughly three times as strong as Earth's magnetic field.

In Chapter 2 we saw how a planet's magnetic field traps charged particles from the Sun and elsewhere. In the case of the Earth, the charged particles create the inner and outer Van Allen belts that range from 400 to 40,000

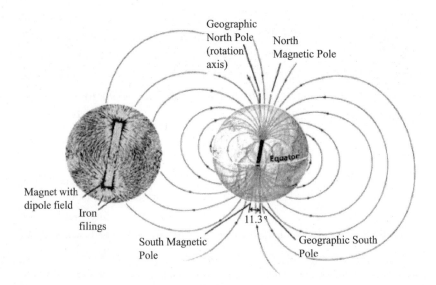

Figure 5.4: A rotating planet with a molten iron core generates a dipole magnetic field, shown here for comparison with the dipole field of a bar magnet. CREDIT: W.H. FREEMAN & CO.

miles above the Earth's surface. Particles trapped in Earth's Van Allen belts have so much energy that spacecraft carrying astronauts to the Moon are sent through them as rapidly as possible so as to avoid radiation poisoning. This poisoning, caused by impacts from high-energy particles, damages many organs in the human body. High doses lead to sickness and death.

The stronger the magnetic fields in a planet's Van Allen belts are, the more particles they can trap. Futura's stronger Van Allen belts will contain many more high-speed particles than the belts around Earth. The danger from radiation poisoning will be dramatically greater when astronauts from Futura venture through the belts than occurs for astronauts from Earth. Indeed, the technology that got astronauts safely through Earth's Van Allen belts might not be sufficient to ensure the safety of Futuran astronauts. Therefore (safe) space exploration by Futurans is likely to be delayed compared to when it occurred here. Likewise, getting robot spacecraft (i.e., those without people) through Futura's Van Allen belts will require higher levels of technology than it does on Earth. Today we have hundreds of satellites in orbits in the Van Allen belts. These must be shielded from the radiation around them. Satellites in Futura's belts will have to have even more protection.

The high-energy particles trapped in the Van Allen belts both here and around Futura are also located at a special distance from Earth where satellites can orbit and remain stationary over one place on Earth (introduced in Chapter 2). Called geosynchronous or geostationary orbits, this is the distance at which the satellites circle the Earth at exactly the same rate that the Earth spins. These satellites are used where continuous communication through them is required. With the higher concentration of potentially damaging, high-energy particles around Futura, these satellites will have to be shielded better than those around Earth, another example of how engineering for space around Futura will have to be more robust than it is here.

Futura's Surface

Let's now consider the consequences of extra heating from inside Futura. Because Futura is composed of more heavy elements than is Earth, with a thicker core than we have, there must be fewer of the lighter elements, including aluminum, silicon, calcium, and sodium, among others, and minerals that are made from such elements. Less of this rocky material implies that both Futura's mantle and crust must be thinner than those of the Earth. The hotter interior of Futura will lead to more convection in the asthenosphere (just the opposite situation to what we saw in Chapter 4). The hot rock will be able to distort and move the thinner lithosphere more easily, implying that there will be more volcanic activity and more active tectonic plate motion there than here.

Futura will clearly be a more challenging place to live upon than is Earth. Although it is difficult to predict how many tectonic plates Futura will have, they will be moving faster than those on Earth due to the higher convection rates on Futura. This will lead to more seismic activity as the lithosphere subducts (one plate going under another), grinds (like the Pacific Plate and North American Plate scraping past each other and creating, among other things, the San Andreas Fault), and separates (in mid-ocean ridges) more rapidly than occurs on Earth. Boundaries of plates, such as the coast of California (if it existed on Futura) would be virtually uninhabitable by people, as the Futuraquakes and volcanoes would occur in these places incessantly.

The overactive volcanism and tectonic plate activity will have an interesting effect, namely bringing more heavy elements and processed elements to the surface of Futura. Shortly after differentiation, a planet's crust has a distinct lack of heavy elements such as iron, nickel, gold, silver, platinum, lead, and many others. The ones found on the surface later on are mostly returned to it from the planet's interior. Because Futura will be more seismically active than Earth, Futura's surface will eventually be richer in these elements. Gold and platinum items will be a dime a dozen.

Equally intriguing will be the more common formation of volcanic pipes in which diamonds form and are transported up to a planet's surface. Called Kimberlite pipes, after the town of Kimberley, South Africa, where one pipe particularly rich in diamonds was discovered, these diamond conveyors will be more common on Futura than they are on Earth. Therefore, the rich deposits of gold and platinum on Futura will be complemented by myriad diamonds. We explore more about Futura's surface shortly.

Futura's Orbit

I have not yet specified how far Futura is from its star. It will not be the same distance that we are from the Sun. Let's see why. We have established that Futura is hotter inside and has a thinner crust than the Earth. In contrast to Dichron in Chapter 4, the heat from inside Futura will be more easily able to escape than can heat from inside the Earth. Where that heat comes out on Futura's ocean bottoms, ocean currents will, for the most part, carry it away along the ocean bottom (i.e., laterally). However, there are places in the oceans where water from deep down has little lateral motion. In such places it moves upward. If this strong vertical motion occurs near heat vents, then warm spots develop on the ocean surface. This especially warm water would heat the air above it, causing that air to rise and begin circulating in ways that would create strong weather patterns, most notably hurricanes and typhoons.

Heat sources that vent directly onto the surface of Futura through volcanoes, for example, will also heat the atmosphere. Heating from both the oceans and the land will therefore lead to the formation of more storms on Futura than are similarly generated on Earth. Combining the heat sources emerging from Futura, we see that more heat reaches its surface, and hence its atmosphere, than comes into our air from inside the Earth.

To compensate for the extra heating it experiences, I put Futura in orbit slightly farther from its star than we are from the Sun. Specifically, to lower the average temperature by 10°F, I move Futura about 3.5 percent farther out. This changes the time it takes Futura to orbit, that is, changes its year, from 365¼ days, as it is here, to 385⅓ days.

LIFE ON FUTURA

Acknowledging the incredible complexity of world-making, I cannot see any fundamental impediments to the evolution of life on Futura. It will have on its surface liquid water, carbon, and other elements necessary for life as we know it, as well as be in orbit around a star that provides the necessary energy for life over billions of years. Its surface will be moderately stable, its atmosphere will be transformable into one composed primarily of nitrogen-oxygen, and it will have a substantial (Moonlike) moon to create and stir up the primordial soup. Let us assume that it tracks similarly to how life formed and began evolving on Earth. Life begins in Futura's oceans and spreads onto the continents.

Aliens

While "normal" evolution is under way there, however, I wish to explore the consequences of an event that is much more likely to occur there, in the far future, than was ever likely to occur here. I believe that by the time Futura is habitable, some $4\frac{1}{2}$ billion years after it forms ($19\frac{1}{2}$ billion years from today), the Musketeer Galaxy will be colonized by one or more races of sentient creatures or, as we would call them, aliens. The event in question is that one of those races sends colonists to Futura.

I assume that the visit by aliens occurs before the first sentient creatures evolve from simpler species on Futura. In other words these visitors from space colonize Futura when it is full of vegetation and animal life, but before it is necessary for the aliens to compete with indigenous people. Let's consider some of the factors that go into such alien colonization.

There are many as yet unknown factors in determining how far from their homes alien civilizations can spread, along with issues surrounding the separation between their colonies and how quickly they can move between habitats, so I make some assumptions and see where they lead us.

Alien Travel Speed

Despite the appeal of *Star Trek*, I assume that the speed of light remains the ultimate speed limit. Einstein's 1905 theory of special relativity predicts, and experiments verify, that the faster one travels, the more mass one acquires. For example, if you are at rest on the Earth, then your mass is what you would measure it to be based on your weight.* I have a mass on Earth of 5.9 slugs or, if you are British, 13.5 stone. Traveling past Earth at half the speed of light, your mass would be fifteen percent greater than your mass on Earth. My mass would be 6.79 slugs. Traveling at ninety-nine percent the speed of light, your mass would increase to an impressive 7.1 times your

---

* Weight and mass are related on Earth by the simple relationship: weight equals mass times the acceleration of gravity (32 ft/sec/sec). The corresponding mass is called a *slug*. To find your mass in slugs, weigh yourself in pounds and divide that number by 32. A *stone* is the mass corresponding to 14 pounds. In the more cultured MKS system of units, weigh yourself in Newtons, divide by 9.8 (dividing by 10 is close enough) to get your mass in kilograms. Indeed, many scales in countries that use the latter system give the results directly in kilograms.

mass on Earth. My mass would be 41.8 slugs. If you were to travel at the speed of light, you would have more mass than everything else in the universe combined!

Because your mass rises so dramatically as you approach the speed of light, it would take rockets of mind-boggling power to accelerate a reasonably sized interstellar spacecraft to even ten percent the speed of light. I therefore assume that our alien colonizers travel at one percent the speed of light.

Separation Between Habitable Worlds

The next assumption to make is the separation between habitable worlds in the Musketeer. Based on the distribution of stars in our neighborhood, it is likely that separation will be at least 1,000 light-years. This means that light takes 1,000 years to go from one habitable world to another in the Musketeer, and also that it takes travelers going at one percent the speed of light 100,000 Earth years to go from one such world to another. This is one of the reasons it is exceedingly unlikely that aliens have visited Earth. In order to get here from a planet orbiting a star at a similar distance in our Milky Way, they would have to travel for many generations if special relativity is correct (and every experiment testing it implies that it is), and if a generation of alien life is similar in length to ours. I can't imagine them popping over to Earth for a quick mission to take blood samples from an unsuspecting human or two and then leaving. Likewise, it is unlikely in the extreme that they are living among us, for reasons that become clear shortly.

It is likely that the trip to Futura will be undertaken by an armada of spacecraft to help ensure that some of the vehicles and the "people" in them will arrive alive, given the challenges of space travel.* This is analogous to the efforts of early ocean sailors such as Columbus and Magellan, who took several craft on each voyage.

If more than one alien species populates a galaxy, it is likely that there would be intervals of hundreds of thousands of years between visits by different species of aliens to a newly discovered habitable planet, such as Futura.

---

* See, for example, the author's *Hazards of Space Travel* (New York: Villard, 2007).

The interval between when travelers from the same race visit an outpost such as Futura is similarly difficult to determine.* For the sake of exploring substantial consequences of colonizing, I am going to assume that one set of colonists settles on Futura and is left alone there for a hundred thousand years. This would allow them to have about 4,000 generations (at twenty-five years per generation) of their species on this world. How would the colonists have to adapt to life on Futura and how might evolution change them over those 4,000 generations? Indeed, would they be recognizable to members of their race arriving a hundred thousand years later?

Adaptation/Evolution

In exploring issues surrounding adaptation, there are some assumptions that we need to make about the colonists and about Futura. First, the colonists are carbon-based life-forms. This was justified in Chapter 2. Second, although Futura will have rocks, sand, oceans, continents, and other structures similar in chemistry and structure to those on the colonists' home world, Futura is not identical to their home world in any way. I assume that Futura is potentially inhabitable by them, but as we show, the colonists will have to use their technology to the utmost, including genetic engineering of the life they bring with them in order to make a long-term go of living there.

Finally, given that the colonists live for 100 years, say, and the trip to Futura takes 100,000 years, clearly the generations that arrive at Futura will have no personal experience with life on their home planet. To help make the potential differences between the two worlds as compelling as possible, let's assume that the interstellar spacecraft that carry the colonists are designed to re-create all the aspects of life on their home world associated with their evolutionary properties. For example, the day–night cycle, as well as daily and annual temperature and humidity cycles on the ship, will be the same as their race experiences at home. The discussion below addresses some of the hurdles that the colonists will have to overcome before they are fully adapted to life on Futura.

---

* Another of the many imponderables is how long any advanced (in this case meaning "spacefaring") civilization survives before war, economics, disease, or any number of other disasters prevents it from being able to support an interstellar fleet.

Acclimatizing to Futura's Atmosphere

The interior and chemistry of Futura are significantly different from those of Earth (using our planet as the archetype of the world from which the aliens departed), but I assume that the atmospheric properties of Futura are sufficiently close to those on Earth that we can meaningfully compare the properties of our air with those of Futura's. One fundamental similarity between the air on Futura and the air of the colonists' home is that both are primarily composed of nitrogen and oxygen. The oxygen is used in the colonists' bodies and they return carbon dioxide to the atmosphere as a waste product. However, as with Dichron in the previous chapter, there is no guarantee that the concentration of oxygen or the total pressure of the air on Futura will be the same as on their home world. I choose to make the total pressure of Futura's air fifty percent higher than the density of the air we breathe and the concentration of oxygen fifteen percent higher. These differences also require adjustments to the rate and depth at which the colonists breathe.

Another difference in atmospheres between Futura and the aliens' home is in the trace gases found in them. The three most common gases in Earth's air, nitrogen, oxygen, and argon (an inert element), comprise over 99.96 percent of its dry volume. Water is found in varying amounts. The rest is composed of trace gases including: carbon dioxide, neon, helium, methane, nitrous oxide, carbon monoxide, ammonia, and ozone, among others.

Life on Earth is adapted to these trace elements in our air. Indeed, the most common of these, carbon dioxide (0.038 percent of the air's volume), is essential for plant life here. The ability of animals to get rid of carbon dioxide is determined, in part, by the difference in the concentration of carbon dioxide in the air and in the animal's lungs. If the external concentration is lower than in the lungs, the carbon dioxide goes from the lungs to the air. If, however, the carbon dioxide is higher in the air than in our blood, then we can't rid ourselves of the CO_2, causing respiratory acidosis. This is the condition when excess carbon dioxide makes the blood more acidic, leading to a whole bunch of really bad things, including coma and death. So, if the colonists on Futura come from a world whose atmosphere has a significantly lower CO_2 concentration than on Futura, they may have trouble exhaling that waste product.

Likewise, if their bodies are not adapted to the other trace elements in Futura's air, then they will have to assist themselves with filters or other technologies. This is one of many examples where evolution will help future generations of colonists. Those first explorers with body chemistries even just slightly better adapted to the air on Futura will have an edge on surviving compared to their companions with more marginal body chemistries. The former are likely to reproduce more proficiently and pass on the genes that, depending on how evolution actually works,* will eventually lead to colonists who are adapted to the carbon dioxide concentration and other trace gas levels in Futura's air.

Assuming the colonists have evolved the abilities to speak and hear, they will have to adapt to different pitches in the sounds they hear because their bodies are adapted to a different air chemistry (percentages of nitrogen and oxygen) than that of Futura. Such variations create differences in the speed of sound and hence the frequencies that one generates and hears. As an extreme example, think of the sound Mickey Mouse makes, which is what you would hear if you spoke through a helium-rich atmosphere.

Adaptation to Their Weight on Futura

Of all the planets in our solar system with solid surfaces (Mercury, Venus, Earth, and Mars) Earth has the strongest force of gravity at its surface. The strength of the force of gravity on a planet's surface is called its surface gravity. In turn, the force of gravity that a planet exerts on you as you stand on it determines your weight. You would feel lighter on the other three planets. For example, I weigh 189 pounds on Earth. I would weigh 167 pounds on Venus, and interestingly enough, 70 pounds on both Mars and on Mercury. The fact that the last two weights are the same is strictly an accident of the fact that Mars is larger, but less dense than Mercury.

The Mars–Mercury coincidence indicates that it is possible that colonists on Futura will come from a planet on which they would have weighed the same as they do on Futura. That, however, is extremely unlikely. Let's assume that the surface gravity on Futura is ten percent higher than on their home

---

* For example, is it punctuated equilibrium, gradualism, a combination of the two, or some other model?

world, meaning that the colonists are ten percent heavier on Futura than they would be back home. Arriving on Futura, the colonists will have to adapt to their greater weight by developing different distributions of muscles than their ancestors had. Their leg muscles, of course, would have to become stronger, but so, too, would the muscles that hold them upright.

Adaptation to Water Chemistry

There are more heavy elements on Futura than on Earth. This is true both inside it and, as a result of the dynamic mantle bringing heavy elements back up to the surface, on the crust. As a result, rainwater will dissolve more heavy elements as it washes down into rivers and then into oceans. Therefore, the chemistry of the oceans on Futura will be very different from the chemistry of Earth's oceans. Indeed, it is likely that virtually all the concentrations of minerals (such as sodium, chlorine, sulfur, calcium, magnesium, and potassium) and the dissolved heavy elements on Futura would require early life evolving there to adapt to a very different water chemistry than did life evolving in Earth's young oceans.

Likewise, the chemistry of Futura's oceans and those of the planet from which the settlers came are likely to be different from each other. We know from experience that it can be dangerous for us to drink water that has higher than "normal" concentrations of metals, such as iron, lead, cadmium, nickel, zinc, chromium, mercury, and selenium, among others. Settlers on Futura will need to process the water there, both from rivers and oceans, before drinking it or even letting it come into contact with their skin.

Adaptation to Bacteria and Viruses

Perhaps the greatest challenge facing the colonists on Futura is adapting to the pre-existing microscopic life that exists there. This has also been a huge problem on Earth when humans go from one region of our planet to another.

The colonists on Futura are likely to be wiped out by germs or viruses endemic to that planet if they are not careful. Identifying the microscopic life-forms on Futura and determining their genetic structures and their potential effects on the colonists will be crucial steps even before they leave their spacecraft. Knowing what can make them ill and even kill them will enable a suitably advanced civilization to develop vaccines and other forms of protection before the problems get out of hand. The ability to develop vaccines on Earth was first demonstrated in 1798 when Edward Jenner developed one against smallpox. That kind of biological research slowly accelerated during the following centuries, with at least five major vaccines in the nineteenth century and more than twenty-five in the twentieth century.*

Nevertheless, until the past few decades, the idea of developing vaccines (and many of them) in matters of weeks or months after identifying a new bacterium or virus would have been "pie in the sky." It used to take years or decades. But now biologists, biophysicists, geneticists, and other scientists are making great strides in rapid identification of dangerous organisms and rapid synthesis of suitable vaccines and cures for people who are infected.

Adaptation to Food Sources

Unlike astronauts who will be traveling to other worlds this century, the colonists on Futura will not be bringing all the food they and their descendants ever need with them. The colonists will have to eat food grown on Futura. Let's consider food from plants first. There will be two options: bring seeds to grow foods they know are suitable for them to eat or modify the plants on Futura to make them edible.

Bringing seeds on trips through interstellar space will expose the genetic information in the seeds to almost certain destruction from high-energy radiation out there. Keeping the seeds in extremely well-protected containers will be absolutely essential, but there are no guarantees that even that will be enough, given the penetrating power of gamma rays and some cosmic rays. Alternatively, given that we are talking about really advanced civilizations, they could store the genetic information necessary to rebuild the seeds on computer media. If all goes well, upon arriving at Futura, seeds could be reconstructed, sown, and, as is very likely to be necessary, genetically modified to adapt to the Futuran soil. Of course, the databank storing the genetic information could also be damaged by radiation en route.

---

* These are for individual diseases. The number is much higher if one counts different vaccines for the same illness.

As a food source, native plants present a variety of challenges to the colonists. First, plants on Futura may contain components that are irritating or deadly to the aliens settling there. Even on their home planet, some plants are likely to be hazardous. Analogously, a variety of mushrooms, belladonna, foxglove, and many other plants contain compounds that are poisonous to humans.

Second, the plants may not contain all the nutrients the colonists need. Consider, as one example among myriad others on Earth, the need for amino acids. These molecules play numerous essential roles in biology, such as participating in synthesizing proteins and other biological molecules and participating in energy-generating processes. Some amino acids can be synthesized in the human body, whereas others must be taken in whole from food. If Futura couldn't provide these latter ones, colonists needing them would have to manufacture them.

To make matters even more challenging in the realm of amino acids, there are two fundamental types of these molecules, called left-handed and right-handed, that are mirror images of each other, like a person's left and right hands. For reasons that are still being explored, nearly all life on Earth is based on one family of amino acids, the left-handed ones. The properties of the left- and right-handed amino acids are so different from each other that eating right-handed versions of essential amino acids that the body takes in whole will not provide you with useful nutrients. In other words, your body can't use the right-handed versions.

What if the colonists on Futura evolved on a world where life was based on right-handed amino acids and Futura's life is based on left-handed amino acids? The colonists would not be able to survive just on native food sources.

The challenge to eating meat sources is equally complex. Besides such issues as the handedness of amino acids, the animal life on Futura would have evolved with the ability to consume and process a variety of heavy metals that are toxic to humans and to the colonists on Futura. As a result, the meat from these animals is likely to have much higher heavy metal content than would be safe for the colonists.

If embryos of animals were brought to Futura to populate that world with food sources, these creatures would face the same heavy metal threats that the colonists face. Whether the animals could withstand the heavy metals they would accumulate from eating indigenous plant and/or animal life on Futura is an open question. In any event, all life brought to Futura from elsewhere that is capable of surviving on that planet would find it a struggle, as they did not evolve to adapt to it. It is virtually certain that their descendants would have to evolve in order to become better acclimatized.

Adaptation to Day and Night Cycles, Annual Cycles, Climate/Seasons

As we saw in Chapter 2, animals and plants on Earth have biological clocks that help synch their activities to the length of the day. If the day–night cycle is different from twenty-four hours by more than about three hours, these clocks, and their owners, are unable to function for long. This leads to death in simpler life-forms and serious disorientation in humans.

It is unlikely that the length of Futura's day will be exactly the same as that of the colonists' home planet. For the sake of their mental health, I assume that the difference between those lengths of time is short enough that they and any other life-forms they bring with them can function on Futura.

Likewise, the length of a year on Futura and the tilt of that planet's axis, both of which combine to determine its seasons, need not be the same as they were back on the colonists' home world. The length of a year is determined by the mass of the star around which a planet orbits and the distance between the star and that planet. Assuming that Futura's axis is tilted more than that of the home planet and that Futura's year is longer, then the colonists will have to adapt to greater changes in temperature, humidity, and intervals of light and dark throughout the seasons, as well as to longer periods for each season.

Adaptation to Existing Animal Life

Consider one final adaptation that the colonists will have to make, namely to the behaviors of the indigenous animals on Futura. Given the incredible diversity of animals on Earth and the fact that most have a variety of similar behaviors, such as protecting and nurturing young, mating rituals, herding, migrating, and establishing and maintaining territories, it is likely that analogous behaviors will occur for animals on Futura.

When the colonists arrive, they will have to break some of these cycles.

As with humans and the other animals on Earth, the colonists on Futura will start taking territory away from indigenous species, as well as forcing changes in migratory routes. As noted earlier, it remains to be seen whether animals that evolved on Futura can be eaten by the colonists. If so, that will create a set of changes similar to what has occurred on Earth, namely hunting, herding, and large-scale slaughtering. If there exist animals on Futura that are so ferocious they cannot be tolerated, such as Tyrannosauroidea, they might have to be exterminated or limited to islands that the colonists avoid.

If all native animals are inedible, then there will be a different set of relationships between them and the colonists. Although hunting to maintain tolerable herd levels will have to occur, the animals will plausibly be left alone as long as they don't bother the settlers. Whether it is possible to domesticate some animals on Futura is an open question.

Evolution

I have given a few examples of where evolution of colonists on Futura would be helpful to them. There are many, many other realms of their existence that would benefit by refined genetic adaptation. Furthermore, as they start changing Futura to suit their needs, the plant and animal life there will undoubtedly evolve in response. Hundreds, perhaps thousands, of generations later the two sets of life-forms, one that began on Futura, the other from a planet a hundred light-years away, will become one, and it would be just as hard for the descendants of the colonists to go back to their home world as it was for the original space travelers to settle on Futura in the first place.

VIEW OF THE UNIVERSE FROM FUTURA

Astronomers on Futura will someday write about the distribution of stars in their own galaxy, the distribution of galaxies throughout the universe, and how the galaxies and the universe as a whole are evolving. Being in a completely different galaxy than we are, the patterns of stars in the night sky that people on Futura will see will be totally different than those we see from Earth. No Big Dipper or Orion from Futura, but there will be other patterns of stars that people there are likely to associate with things in their everyday lives.

By plotting the locations of other galaxies, they will discover that galaxies are grouped together in clusters. These clusters are also clustered together, in superclusters, which are the largest groupings of objects in the universe. The superclusters are located on boundaries, like the surfaces of bubbly soap or the fibers of a sponge, separated by monstrous regions with very little matter (i.e., stars, gas, or dust) in them. These relatively empty spaces are called voids.

There will be some profound differences between the numbers and locations of galaxies, clusters of galaxies, and superclusters of galaxies between what we see now and what Futurans will observe.

Clusters of Galaxies

Today, galaxies are found grouped together in gravitationally bound clusters of between roughly 10 and 1,000 galaxies. These galaxies orbit each other. Sometimes they collide and scatter each others' stars into intergalactic space. Sometimes they collide and merge.* Therefore, by the time Futuran astronomers study clusters of galaxies, they will be seeing fewer galaxies in each one than do astronomers on Earth today.

The percentages of each type of galaxy in the universe will also change from what they are today by the time Futurans make their observations. As we saw earlier in this chapter, galaxies occur in only a few shapes: normal spiral, barred spiral, lenticular, elliptical, and irregular. When the Futurans observe clusters of galaxies, they will see more elliptical galaxies and fewer normal or barred spirals than we see today. In addition to being destroyed by collisions, fewer spirals will exist because many of them will stop maintaining spiral arms. They will become lenticular.

Cosmic Perspective

Not only will the clusters of galaxies change, but so will the scale of the universe. We saw earlier in this chapter how astronomers know that the uni-

---

* For more on these activities, see Chapter 10.

verse is expanding. As it gets larger, we are actually seeing farther away and, therefore, farther back in time. Nineteen and a half billion years from now, Futuran astronomers will be able to see a universe roughly four times larger than what we can see today. This expansion will create two major changes in the universe. First, the voids will grow larger and the superclusters will spread apart. In other words, the universe will look emptier than it does now.

Second, Futuran astronomers will be able to see many more galaxies than we can see today. This may sound contradictory with the statement above that each cluster will have fewer galaxies, but it isn't. Recall that at the beginning of this chapter we noted that by looking farther away in space we are looking farther back in time. Associated with this concept is the fact that the universe is so large that light from most of it has not reached us yet

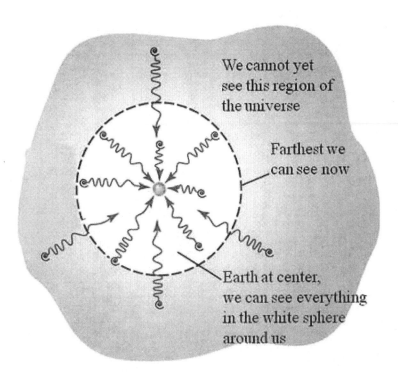

Figure 5.5: *The white region located at the center of this figure illustrates the sphere of space around us. Squiggly lines show how far photons have traveled from the indicated galaxies since the beginning of time. Light from all galaxies and other objects located within the white region has reached us. Light from objects in the gray region emanate from so far away that it has not yet gotten to us—we can't yet see anything beyond the dashed circle. As the universe ages, we see more of it.*

(Figure 5.5). We can only see objects today that are within 13.7 billion light-years of Earth. Those that were formed 20 billion light-years away, say, won't be visible for another 6.3 billion years. At that time, the galaxies 20 billion light-years away will appear to us as they did when they were just forming, 20 billion years ago.*

In fact, Futura will develop in a universe that is using up all of its raw material, namely the hydrogen and helium formed shortly after the Big Bang. Unless there is another set of events about which we astronomers presently know nothing, as time proceeds, once that fuel to make stars shine is gone, it is gone. Then the number of new-formed stars will decrease and the universe will expand and grow dimmer.

---

* Actually, the acceleration of the universe, discussed earlier in the chapter, will spread it out more than this, causing it to look even more empty.

6 What If There Were a Counter-Earth, a Planet in Earth's Orbit on the Other Side of the Sun?

DOPPLE I, II, AND III

"Where are they hiding?" Admiral Baxter demanded.

"There's a whole lot of space out there, sir," Lieutenant Manarino replied, studying the three computer screens in front of him.

"I know that, Mr. Manarino. I also know that we should have more than enough technology out there to find them." He sighed. "I know it's not your fault, son. But I am not going to pay ransom for the governor of the Moon. We are going to find whoever took her and we are going to take her back."

"I tracked them into space and out of lunar orbit, sir, but they vanished and it's been sixteen weeks. They could be anywhere from here to Jupiter."

"We've manually searched Mars and all the moons, asteroids, and comets within range?"

"Yes, sir."

"You have people figuring out how they 'disappeared'?" he asked with the slightest hint of derision.

"Oh, yes sir. We know how they did it. They took the standard battlefield cloaking device and reconfigured it to match all the frequencies that we use for space radar."

"Even allowing for Doppler shifts, because they must have been riding their rockets for part of the time?"

"Clearly they took that into account, sir."

"Now that you know how they did it, I assume our people have devised a different surveillance system they can't hide from?"

"Yes, sir. We built it and tested it every way till Tuesday."

"And?"

"Nothing, sir. We can't find them. As you know, we have a catalogue of the locations of every human-made object in space. We found all of them with this new system. Everything else up there is a natural object."

"Obviously not, because we still don't know where they are. Listen, son. You know as well as I do that until we find them, we don't know if the new radar really works."

Lieutenant Manarino also knew that silence was appropriate. He waited. Admiral Baxter had been a solar system–renowned physicist in his prime, responsible for several key components in the current generation of ultra-high-thrust interplanetary rockets. He was legendary in not taking credit for his achievements, being the first person in history to turn down the Nobel Prize.

The admiral left and walked, distractedly, for half an hour through the corridors of the Pentagon. Then he brightened and strode purposefully back to his offices.

"Mr. Manarino, I believe I know where they are hiding."

"Sir?"

"Stable Lagrange point."

"Sorry, sir. I'm not familiar with the concept."

"Because of the gravitational field created by each planet, there are two places in the solar system where the gravitational force of each planet plus the gravitational force of the Sun combine to hold tiny objects, space debris usually, in orbit. If something like an asteroid starts drifting away from a stable Lagrange point, the gravitational forces from the planet and Sun pull it back. I'll bet they are using gravity to hide from us. If they put their spacecraft at one of Jupiter's Lagrange points and powered it down so we couldn't see an electromagnetic

signature they could stay there for months or more and never be detected. They might even be tethered to a Trojan asteroid.

"Have we sent missions to the Trojan asteroids?"

"Um, no sir."

"Order them out."

I remember as a teen watching a movie in which people traveled to an inhabitable planet on the opposite side of the Sun from us. The prem- ise was that it was always hidden from Earth so it was only discovered when space travel allowed us to view it directly. That world was, of course, inhabited by people who looked just like us but everything there was a mirror image of things here. For example, they wrote everything back- wards.

The belief that there is a "counter-Earth" in the solar system extends back at least 2,500 years to the work of the Greek philosopher Philolaus (ca. 470 BCE–ca. 385 BCE). His convoluted ideas, derived before the concept of the gravitational force was known, had the Earth and the counter-Earth orbiting a central fire of the universe, which was not the Sun. Therefore, Philolaus's counter-Earth was not located on the opposite side of the Sun from the Earth.

With the discovery that the Earth orbits the Sun, counter-Earths have been proposed on the opposite side of it from us. Could we ever see it from here? At first glance, that seems impossible because it is on the "opposite" side of the Sun. But there is a subtlety we need to consider related to the shape of Earth's orbit around the Sun and the fact that our orbital speed is not constant. We go around the Sun in a slightly elliptical (egg-shaped) orbit, which has the prop- erty that the closer we are to the Sun, the faster we are moving, and vice versa. Such orbits apply to everything orbiting the Sun, including the hypothetical counter-Earth.

Let's assume for the moment that such a world is identical to Earth and that the two bodies are in exactly the same orbit. When Earth is closest to the Sun the counter-Earth is farthest from it (Figure 6.1). According to Kepler's second law of orbital motion, "A line from the center of the Sun to the center of a planet sweeps out equal areas in equal times," we deduce that at that time Earth is moving fastest, while counter-Earth is moving slowest. Therefore, we

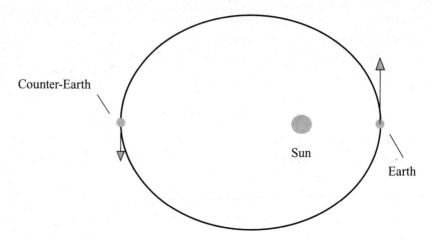

Figure 6.1: Orbits of Earth and counter-Earth. Kepler's laws reveal that at the time indicated, Earth is moving fastest in its orbit and counter-Earth is moving slowest. Arrow lengths indicate relative speeds. The ellipticity ("oval-ness") is exaggerated for clarity.

will slide around the Sun faster than it does and moments after the initial alignment, the Earth, the Sun, and counter-Earth will no longer be in a straight line. If there is any time when we might see the counter-Earth, this is it. However, covering a disk half a degree across, the Sun is so large in our sky that even when the two versions of Earth are out of line with it, the Sun still blocks counter-Earth from our view. So we would never see the counter-Earth from here.

We do know that it is not there, however, because in the twentieth century we sent spacecraft into orbits where they could see it if it were. They didn't. But just because it isn't there doesn't mean that it couldn't be. In this chapter we explore the possibility of whether a world I call Dopple (not necessarily a twin to the Earth) could exist in the same orbit as the Earth on the opposite side of the Sun from us.

FORMATION OF DOPPLE

Planets, we saw in Chapters 1 and 4, coalesce from tiny pieces of space debris orbiting in a disk around a young star. For concreteness, consider our Sun and its disk. Following innumerable collisions of space rubble, eight protoplanets grew larger than their neighboring debris in orbit around the Sun.

Smaller pieces of rock, metal, and ice were pulled onto these protoplanets, which grew into the terrestrial planets and into the cores of the giant planets Jupiter, Saturn, Uranus, and Neptune. This scenario is likely to have occurred millions, perhaps billions, of times throughout our galaxy and others.

Staying with the Earth and solar system for this chapter, let's assume that a ninth protoplanet destined to become Dopple actually formed on exactly the opposite side of the Sun from the Earth. There are three possibilities for Dopple's evolution.

- If there weren't as much debris for it to grab as occurred for Earth, Dopple would have forever remained smaller than the Earth.
- If there were a roughly equivalent amount of nearby debris, Dopple could have grown to a size and mass similar to Earth.
- If there were more debris on its side, Dopple could have become much more massive than Earth. Indeed, we have now seen Jupiter-like gas giant planets around other stars at the same distance from their stars as we are from the Sun. Dopple could be such a giant.

For reasons that become obvious momentarily, I reserve the right to set Dopple's distance from the Sun slightly farther or slightly closer than that of the Earth. Let's consider the three possibilities for Dopple.

Counter-Earth Much Less Massive Than Earth—Dopple I

In order to determine how far from the Sun to place this version of counter-Earth (Dopple I), we need to know whether two objects with different masses but in identical orbits take the same time to go around the Sun. The answer comes from Kepler's third law, which relates the period of an orbit around the Sun (i.e., the length of a year) to the mass and distance of the body from the Sun. We astronomers compare all orbits in the solar system to Earth's year and our planet's average distance from the Sun, called an astronomical unit (AU, for short). What Kepler's third law reveals is that if Dopple I had exactly the same orbit as the Earth, it would take a slightly longer time to go around the Sun than does the Earth. This tiny difference would add up, causing Dopple I to drift away from its location opposite us and therefore become visible from Earth. To compensate, I put

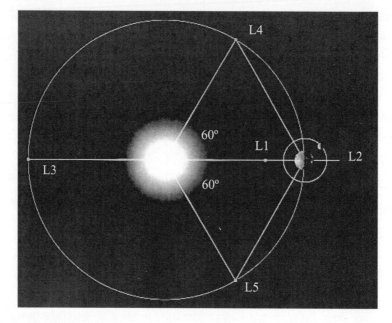

Figure 6.2: The five Lagrange points where the gravitational forces of the Sun and a planet combine to create either stable (L4, L5) or unstable (L1, L2, L3) orbits. COURTESY: NASA/ JPL- CALTECH.

Dopple I slightly closer to the Sun than 1 AU so that its year exactly equals that of the Earth.

Will Dopple I stay opposite the Earth? The answer is revealed by a careful analysis of the gravitational effects of the Sun, Earth, and other planets in the solar system on Dopple I. The calculations were first done in 1772 by the Italian–French mathematician Joseph-Louis Lagrange. He discovered that there are five special points related to a planet's orbit around the Sun, including the one we are considering (Figure 6.2). These are now called Lagrange points. Any object that is much less massive than the planet under study and that is located at any of these five Lagrange points feels gravitational forces that greatly affect its orbit. A quick aside on the Lagrange points is worthwhile.

Lagrange Points

The Lagrange points are the locations where the force of gravity from the Sun and force of gravity from a planet combine to attract or repel a low-mass object that is located at them. The gravitational effects at L4 and L5 are easiest to understand. These two points are located roughly sixty degrees ahead of the planet in its orbit around the Sun and sixty degrees behind it.

Imagine that there was a small asteroid; say ten miles across, orbiting at L4, sixty degrees ahead of Earth. Because the Earth and the asteroid orbit faster than the outer planets, eventually the asteroid will pass between Jupiter and the Sun. The tug of the giant planet would pull the asteroid slightly outward from its otherwise elliptical orbit around the Sun. Does the asteroid continue drifting away from L4 after the gravitational influence of Jupiter dwindles? In fact, it doesn't. As soon as the asteroid starts moving away from L4, the gravitational attractions of the Earth and Sun begin pulling it back. This is analogous to the behavior of a marble on a roulette wheel if you were to carve a little bowl-shaped valley in the wheel that captured the marble as it rolled around. Once in the valley, slight tugs or pushes on the marble would cause it to rattle around in the valley, but it would take a major shove to get it out. The same analysis applies when the asteroid is orbiting at L5, sixty degrees behind Earth. L4 and L5 are therefore called stable Lagrange points.

The effects that occur at the remaining Lagrange points, L1, L2, and L3, are more subtle. They are called metastable positions. The combined gravitational forces of the Sun and planet allow objects orbiting the Sun at these locations to stay at them unless they experience even the slightest gravitational pull from another object. When that occurs, the object at one of these Lagrange points will feel pulls from the Sun and planet that actually accelerate it away from the Lagrange point. This is analogous to the situation you would get by taking a large plastic sphere, cutting it in half, setting one piece with its flat side on a table, and putting a marble at the very top of it. The slightest disturbance will cause the marble to roll off. The top of the sphere is a metastable location for the marble.

We have seen that for Dopple I to orbit the Sun at the same rate as does Earth (once an Earth year), Dopple I (with much lower mass than the Earth) has to be in a slightly different orbit from our planet. That result was determined assuming that the Sun was the only body whose gravitational force was acting on Dopple I. Now let's add the gravitational effect of the Earth, which is located on the opposite side of the Sun from Dopple I. Calculations show that including the Earth's effects on Dopple I, that body will initially orbit opposite Earth at our L3.

If Dopple I were initially at the metastable L3, and if the Sun and Earth were the only bodies acting on it, then Dopple I would remain on the opposite side of the Sun from Earth. The catch is that there are gravitational effects from the other planets acting to topple Dopple I.

As soon as the gravitational pull of any object causes Dopple I to slide even inches away from L3, it begins on a slippery slope from which there is no recovery. Pulled away from L3 by the gravitational pull of a third body, the gravitational pulls of the Sun and Earth conspire to pull it even farther away from that location. As a result, Dopple I will be pulled into an orbit that is not exactly equal and opposite to that of the Earth. Orbiting the Sun at a different rate than the Earth, Dopple I will begin drifting toward us. If it is orbiting faster than Earth, it will approach us from behind. If it is orbiting more slowly, we will close in on it.

Clearly, putting Dopple I at precisely Earth's L3 is a bad move. What if we ignore Earth's gravitational effect and put Dopple I in the previously mentioned orbit that gives it exactly the same year we experience? It turns out that this location is so close to L3 that the gravitational pulls of the Sun and Earth will cause it to drift away from that location, just as it was forced to drift away when it was initially right on L3. In summary, there is no stable orbit for Dopple I on the opposite side of the Sun from us.

Let me defer the fate of Dopple I until after we consider whether Dopple III with much more mass than Earth and Dopple II with about the same mass as our planet could stay on the opposite side of the Sun from us.

Counter-Earth with Much More Mass Than Earth—Dopple III

The scenario of Dopple being much more massive than Earth is especially easy to handle in light of the presentation of Dopple I. Therefore, I want to get the more massive Dopple III out of the way before discussing Dopple II, which has about the same mass as Earth. Dopple III is a Jupiter-like planet that is composed primarily of hydrogen and helium surrounding a solid Earthlike core. To be concrete, let us assume that Dopple III has 318 times the mass of our Earth, which is Jupiter's mass.

Recall that the results related to Lagrange points only apply when two of the bodies are much more massive than the third. This third, lower-mass object is the one that can be located at the Lagrange points associated with the other two bodies.

Clearly, Dopple III does not satisfy the criteria of being a low-mass object compared to the Earth, so Dopple III is not at Earth's L3. However, Earth is a low-mass object compared to both the Sun and Dopple III, so Earth is located at Dopple III's L3! The entire analysis we did for Dopple I applies in reverse to Earth in the present scenario. Therefore, without further ado, let us put Earth at precisely Dopple III's metastable L3. We don't even need the gravitational effects of another planet to knock us off L3; the Moon will do that.

In everyday conversation we talk about the Moon "going around the Earth." As we saw in Chapter 1, this is not correct. The Earth and Moon waltz around their center of mass, the barycenter. It is that point that smoothly orbits the Sun today while the Earth and Moon are going around it. With the Earth at L3, the motion of the Moon pulling the Earth around the barycenter would cause the Earth to slide off L3 by 1,000 miles, which compared to our 93 million miles from the Sun, isn't much. But that slight perturbation from the Moon is enough to ensure that once off L3, the Earth–Moon would continue to move away from it.

An alternative scenario is to put the barycenter on L3, but even this is only metastable. The slightest gravitational tug from other planets will cause the Earth–Moon system to slide away from it. In either of these two cases, with the Earth initially on L3 or the barycenter of the Earth–Moon system on L3, the Earth is going to end up orbiting toward the more massive Dopple III.

Counter-Earth with About the Same Mass as Earth—Dopple II

Our last, best hope for a stable Dopple is a planet with about the same mass as the Earth. For concreteness, let's assume that Dopple II has exactly the same mass as our planet. The discussion here can be generalized to bodies of similar mass. In order for Dopple II to have exactly the same orbital period (i.e., the same length of its year as our year) around the Sun, that planet must have exactly the same orbit around the Sun as does the Earth, with an average distance of 1 AU. As discussed earlier in this chapter, the two planets, Earth and Dopple II, initially have identical, elliptical orbits around the Sun. This quickly changes.

We saw that because the speeds of Earth and Dopple II are different at different places in their orbits, they will not always be on exactly opposite sides of the Sun. During the times that they are off-center, the Earth and Dopple II will pull each other away from their original elliptical orbits. Combining these changes in orbits with the tugs from the other planets, which will not be the same on both Earth and Dopple II at the same time,* these two bodies will be pulled out of their original orbits by different amounts. For example, Dopple II, passing between Jupiter and the Sun, is pulled outward by Jupiter more than Earth is at that time because Earth is then on the opposite side of the Sun from Jupiter.

FATE OF THE DOPPLES

Due to the onslaught of different amounts of gravitational attraction from different planets pulling in patterns that do not reverse or frequently repeat themselves, Dopple II and Earth will drift away from being exactly opposite each other. As with Dopples I and III, this broken symmetry leads to the two planets drifting toward each other, especially if their masses are not identical. This would be the case in the likely scenario that Earth was struck and formed a Moon and Dopple II did not. This brings us to the question of the fates of the counter-Earths.

I start the three Dopples at L3 on the opposite side of the Sun from the Earth. Because this is an unstable point, they will drift away from it. This motion will cause the orbits to change ever so slightly. As a result, they will not orbit the Sun at exactly the same rate as the Earth, meaning that they will slide toward it. There are several possible outcomes as the two planets converge.

As these worlds edge toward Earth, they will pass through one of the two stable Lagrange points, L4 and L5 (see Figure 6.2). Whether they get trapped in that Lagrange point depends on how fast they are moving when

---

* This occurs because the other planets will usually be at different distances from our two planets of interest. Gravitational forces acting at different distances have different strengths.

they near it. It is like what happens when you drive your car toward a dip in the road. If you are going fast enough when you approach it, you can coast down it and back up the other side. If you are not going fast enough, you will coast down, but not be able to get back out without using the engine.

Dopple I

If it is moving fast enough relative to Earth to pass the stable Lagrange point, Dopple I could collide with our planet. The collision of a very low-mass Dopple with Earth would be a mess, but it would not have been at all unusual back in the day. Collisions such as we are considering now took place numerous times during the Hadean period of planetary formation. Let's visualize a collision wherein Dopple I strikes the Earth billions of years ago by falling straight down at our planet: Assume that Dopple I had $\frac{1}{300}$ the mass of Earth. This is roughly a quarter the mass of our Moon. The impact would have devastated the Earth's young, thin crust. Although solid parts of it may have remained, like ice floes in the Arctic, most of the Earth's surface would have become molten again, as it was when the planet first formed.

The energy of impact is not great enough to destroy the Earth or to allow much of the debris that is splashed off to escape the planet's gravitational clutches; most of the matter from the two worlds that gets splashed off will eventually fall back down.

One might think that such an impact, creating the mother of all craters, would leave a permanent scar, but this won't happen. The impact region will quickly fill with lava from inside the earth (similar to the way giant craters on our Moon were filled in, creating the dark regions we call maria). The rest of the signs of the impact will be erased by the Earth's tectonic plate motion. We saw in Chapter 4 that the Earth's crust is in continuous motion as a result of the tectonic plate motion driven by our planet's molten interior. That activity would have erased the scar created by Dopple I's impact billions of years ago.

Another possibility is that if Dopple I is moving slowly enough relative to Earth, but not on a collision course with it, then the Earth could capture Dopple I as a moon. Indeed, this could be an alternative scenario for the capture of the moon Anillo in retrograde orbit, as discussed in Chapter 3.

Dopple III

Similar to Dopple I, Earth could fall onto Dopple III or Earth could go into orbit around it, as explored in Chapter 2. Upon collision, the much more massive Dopple III will swallow little Earth. Recall that Dopple III is similar to Jupiter, which means that this Dopple has an Earthlike core of perhaps thirteen times the Earth's mass surrounded by 305 Earth-masses of liquid hydrogen, helium, and water. The collision will begin with a titanic splash, like throwing a huge boulder into a lake. A substantial amount of Dopple III's outer liquid layers will thereby be ejected violently into space. This ejecta will turn to a fine gas, most of which will be pushed out of the solar system as a result of being struck by radiation and particles from the Sun. This is the same effect as sunlight and particles from the Sun pushing the tails of comets outward. Most of the rock and metal debris splashed off Earth and into space is likely to be pulled back down onto Dopple III.

Earth is likely to shatter on impact, with much of its water gushing into space along with Jupiter's outer layer. The incredibly hot core of our planet, settling inward, would vaporize large volumes of Dopple III's remaining hydrogen and helium. Dopple III, like Jupiter, started with 318 times as much mass as the Earth. It gained the better part of the Earth's mass in the collision, but it is likely to lose at least ten Earth-masses of hydrogen, helium, and water due to the collision and subsequent vaporization. Therefore, after the merger is complete, the resulting single planet will be significantly less massive than Dopple III and, from the perspective of life as we know it, completely uninhabitable.

Dopple II (Equal-Mass Planets)

A pair of equal-mass planets orbiting the same distance from the Sun will not strike each other or even pass near each other. It turns out that the two bodies will stop moving toward or away from each other when they have drifted to positions sixty degrees apart, as measured from the Sun (Figure 6.2). Thereafter, like objects at the two Lagrange points L4 and L5, they will be in stable orbits with respect to each other. Even when they are disturbed by the gravitational tugs of the other bodies in our solar system, the two planets would remain in these orbits.

As seen from Earth, Dopple II would always have the same gibbous phase.

It would be brighter than Mars. Because Venus is closer to the Sun and scatters more light into space than does the Earth, Venus will sometimes be brighter than Dopple II. But when we see mostly the dark side of Venus, Dopple II will be brighter than that planet, too.

When life evolved on Earth, they would definitely see Dopple II and, if the latter planet is sufficiently similar to Earth, when people come to exist on Dopple II, they will see Earth. The existence of two life-supporting worlds essentially side by side would be an ideal opportunity to see how evolution works. If the conditions on the two planets were essentially identical (chemistry and large moon come to mind here), then what would the two sentient races look like? I would wager a year's salary that their differences would be as profound as their similarities.

What If the Earth Had Formed Elsewhere in the Galaxy?

KERN I AND II

THE DISCOVERY IN 2016 OF THE FIRST EXTRATERRESTRIAL CIVILIZATION

The research groups on Earth involved in the search for extraterrestrial life use telescopes sensitive to radio waves and microwaves (which are often classified as especially short-wavelength radio waves) to continually monitor for signals from the vicinity of over 5,000 stars. Their goal is to identify transmissions generated by intelligent life elsewhere in our Milky Way Galaxy. This work began in 1960 with the observations of astronomer Frank Donald Drake, using a telescope at the National Radio Astronomy Observatory in Green Bank, West Virginia, in the United States. He found nothing of note.

Searching for radio signals from advanced life is built on the assumption that they send signals similar to our radio communication using AM (amplitude modulation), FM (frequency modulation), or PM (phase modulation) signals. Such communication might be used by the aliens for sending signals to and from spacecraft, to and from colonies on different worlds, to try and attract the attention of other civilizations, like us, or purely incidentally, such as radio signals intended for communication on the surface of a world that leaked into space.

The tremendous distances involved in receiving signals from worlds orbiting other stars also imply that any radio waves we might receive are beamed, rather than sent in all directions. This is true because unidirectional signals are much more energy efficient than omnidirectional signals. Focusing radio waves poses no technological obstacle. We do it all the time using, for example, phased array antennas or parabolic antennas that have been available since the twentieth century.

The use of directional beaming makes economic sense for the civilization sending the signals, because all the energy is aimed in the direction of interest, such as toward a spaceship or between known worlds. However, this technology makes it more difficult to detect those signals if they weren't intended for you: you have to be in the line of sight of the signal or you won't receive it. This is similar to detecting a laser beam at visible-light wavelengths: if the laser beam isn't aimed at you, you won't see it.

Early searchers for extraterrestrials occasionally found energy spikes at the frequencies they monitored; however, none of these events was repeated. Receiving a signal with the same signature twice or more would have suggested that they were sent intentionally (for whatever reason). On November 23, 2015, a sequence of phase-modulated signals was received from the gamma star in the constellation Microscopium located 230 light-years from Earth. Similar in properties to the Sun, this star had been studied from time to time, but not consistently until after that first burst of nonrandom signals.

A similar set of signals was received starting on November 1, 2016. It lasted about twelve minutes and then was heard again every eighteen hours for three weeks, after which time the signal began to fade and then disappeared completely on November 24, 2016. This cyclic pattern is consistent with signals emitted in a fixed direction on the surface of a planet that is rotating once every eighteen hours and orbiting its star at the same distance as the Earth is from the Sun.

Decoding the signals took about six months, in part because phase modulation is rarely used in our technology and in part because the content and format of the signals were alien to us. One thing that took a while to realize was that the signals were analog, rather than digital, which is standard in our communications technology today. When the signal content was finally decoded, we saw our first images of sentient aliens. The images were from a show analogous to a morning soap opera on Earth.

For thousands of years people believed that the Earth was at the center of, well, something. First, we were at the center of the universe. This was codified in the second century AD by Egyptian polymath Claudius Ptolemaeus (i.e., Ptolemy). Then, with the Copernican revolution of the sixteenth century placing us in orbit around the Sun, Earth lost that elite position. Nevertheless, observations suggested that our solar system is located in the center of the Milky Way Galaxy.

Indeed, the bright band of the Milky Way seems to wrap itself around the solar system, reinforcing the belief that we are at the center of it. In the late eighteenth century the German/British musician, composer, and astronomer Sir Frederick William Herschel (Sir William to his friends) counted stars in different directions around us. He found essentially the same number in each direction, supporting the conclusion that we are in the center of our galaxy. His conclusion was in error because he did not know of, and therefore did not take into account the effects of, interstellar gas and dust on his observations. Like clouds in the sky, that debris prevented him from seeing very far into the galaxy, so his star counts were only for very nearby stars; they did not provide insight into the overall structure of the galaxy or our true place in it.

Herschel's conclusions held sway for nearly a century and a half until about 1918, when American astronomer Harlow Shapley studied large groups of stars called globular clusters. These clusters orbit our galaxy in every direction. Therefore, most are located outside the region of gas and dust that fills the band of light we call the Milky Way.* Herschel, by contrast, had only studied stars in the band of the Milky Way.

Shapley found that the globular clusters are not uniformly distributed

---

* The Milky Way appears as bright as it does because starlight scatters off the gas and dust in it, causing that material to glow.

Figure 7.1: *The solar system is located between spiral arms of the Milky Way about halfway from the nucleus to the galaxy's visible edge. The disk of the Milky Way Galaxy has a bar of stars, gas, and dust crossing the center (the nucleus), two major spiral arms, and several minor arms.*
COURTESY: NASA/JPL-CALTECH/R.HURT (SSC-CALTECH).

around us, but rather there are more of them in the direction of the constellation Sagittarius than anywhere else. He correctly concluded that the globular clusters orbit the center of our galaxy and therefore the center of our galaxy must be in the direction where most of them are seen, namely toward Sagittarius. Furthermore, he calculated that the center is some tens of thousands of light-years from us. Both of these conclusions have withstood the test of countless observations: we are located in the suburbs of our Milky Way Galaxy (Figure 7.1).

Although our recent ancestors discovered that Earth is not in the center of the universe or in the center of the galaxy or even in the center of the solar system, the idea that at least our galaxy is in the center of the universe got a temporary boost early in the twentieth century. In 1929, American astronomer Edwin Powell Hubble discovered that all distant galaxies were moving away from the Milky Way and that the speeds at which they are receding depend just on their distance from us. At first glance, this certainly seems to suggest that we are in the center. We saw in Chapter 5 why this isn't so.

We are not in any "preferred" position in the universe. But perhaps being nowhere special is special. To begin with, we are located between two spiral arms roughly halfway between the center and the outer visible edge of the galaxy's disk (Figure 7.1). For the most part, we have gotten on well with the neighbors, the nearby stars and interstellar clouds of gas and dust visible in the night sky. Occasionally, however, some astronomical event has gotten in our face, incidents from which life on Earth has always suffered. It has been proposed, for example, that especially powerful stellar explosions of nearby stars called supernovae may have been responsible for one or more of the mass extinctions that have occurred over the history of the Earth.

Despite such astronomical events, complex life manages to flourish here. But different regions of the galaxy have distinctly different features and properties. Could we do so well if the solar system were located elsewhere in the galaxy? Could such a richly diverse, life-supporting world as Earth have developed and survived at the center of the galaxy? Or in the bar of stars, gas, and dust that pierces the center of the galaxy? Or above the spiral arms? Or beyond them? Or elsewhere in the disk other than where we are today? The answers appear to be: no, no, no, no, yes. Let's consider each of these, keeping in mind the discussion of the Milky Way presented in Chapter 5.

HOME AT THE CENTER OF OUR GALAXY

We aren't at the center of the galaxy, but could we be? Activity in the center of our galaxy is dominated by a black hole with about four million times the mass of our Sun. It is located in the core, technically the nucleus, of the Milky Way. A black hole is a region wherein the matter has become sufficiently concentrated that its gravitational force literally changes the shape of space so much that matter and most energy that enter it cannot escape. One exception to the rule of nothing escaping is the force of gravity created by the matter and energy inside the black hole, which does extend out into the rest of the universe.

This black hole is not an anomaly. Astronomers are discovering that such "supermassive" black holes, each with between a hundred thousand and a

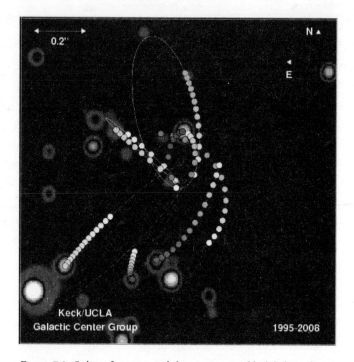

Figure 7.2: Orbits of stars around the supermassive black hole in the nucleus of the Milky Way Galaxy. Each dot represents the position of a star in successive years between 1995 and 2008. The dashed lines show their complete orbits. The black hole is in the center of this image. COURTESY: KECK/UCLA GALACTIC CENTER GROUP.

few billion solar masses, exist in the nuclei of most galaxies. The monumental gravitational attraction of the Milky Way's supermassive black hole is the main arbiter of activity in the galaxy's center. Among the objects we see orbiting it as rapidly as planets orbit the Sun are full-fledged stars (Figure 7.2). Plausibly, planets could form around such stars. But would these planets provide stable platforms on which life could evolve?

Contrary to common belief that black holes are giant vacuum cleaners in space that suck everything into them, weird effects only occur within a few radii of these objects. For example, if a black hole is a million miles in diameter, then bizarre behavior of space, time, matter, and energy occur only within ten million miles of the black hole, which is a tiny distance on the scale of a galaxy. Earth, for example, is ninety-three million miles from the Sun. At the distances that the stars closest to it orbit the Milky Way's supermassive black hole as shown in Figure 7.2, they only feel a normal gravitational pull from it.

Unlike the planets in our solar system, which all orbit in essentially one plane around the Sun, the stars orbit in all directions or planes around the galaxy's supermassive black hole. They have very elliptical orbits around that black hole, meaning that throughout their orbits their distances from it vary tremendously. It is not unusual for these stars to be ten times farther away at the most distant points in their orbit compared to where they are closest to the black hole.

The change in distance between the black hole and a star means that the orbits of their planets will also be continually changing due to the changing gravitational attraction of the black hole. Likewise, the other stars in close orbit around the black hole will pull on the planets orbiting neighboring stars. It is therefore unlikely in the extreme that any planets around stars in the nucleus of the galaxy will orbit reliably in the Habitable Zone of their star. Even planets whose average distance from their star is in the Zone would sometimes be in regions that are too hot or too cold for water to remain liquid on their surfaces (see Chapter 2). Plausibly, a passing star could rip the planet from another star completely out of orbit.*

Even farther out in the galactic nucleus, beyond where stars are in tight orbits around the black hole, planets will experience effects that are very likely to keep them uninhabitable. It is likely that all stars in this region would be on such chaotic orbits that they would frequently pass close enough to each other to disturb comet clouds surrounding neighboring stars.

Recall from Chapter 2 that the solar system, and from what astronomers are observing, most other stars, are surrounded by spherical distributions of perhaps trillions of comet nuclei and related objects called Oort comet clouds. The objects in these Oort clouds are frozen bodies composed of ices, rock, and metal. In our solar system they orbit much farther from the Sun than Neptune. Although we don't know the extent of this cloud of comets, it may extend out from the Sun for a light-year, which is roughly 66,000 times farther than the Earth is from it. Other comet nuclei orbit between the Oort cloud and about Neptune's orbit in the Kuiper belt.

Most of the Oort cloud and Kuiper belt objects have orbits that will never take them into the region where the planets orbit the Sun. However, when

---

* It is possible, although highly unlikely, that two stars would collide; there is so much space between them, even in such crowded astronomical environments as we are discussing here.

stars pass within a light-year or two of each other, their gravitational attractions disturb their mutual Oort clouds and Kuiper belts, causing many of the comet nuclei to assume orbits that do bring them into the realm of the planets. After such events, the planets in these systems are pounded by impacts that can lead to mass extinctions on inhabited worlds. Although such events may have occurred occasionally on Earth, they would happen frequently if the solar system formed in the center of the galaxy.*

Another profound problem facing life-supporting planets in the central region of the galaxy comes from the discovery that clusters of stars have formed there within the past few million years. This is very recent in astronomical terms, suggesting that new stars have formed there from time to time over the billions of years that the galaxy has existed.† Among the new stars in the nucleus are some very massive ones. These are the stars that explode as supernovae within a few tens of millions of years after they first form. As we discussed in Chapter 5, supernovae are titanic explosions in which both matter, in the form of the outer layers of massive stars, and lethal radiation, in the form of gamma rays, x-rays, and ultraviolet radiation, are expelled into space. Life on planets within perhaps a hundred light-years of these supernova will suffer profound trauma including: the elimination of the ozone layer that protects life from ultraviolet radiation; the loss of significant amounts of atmospheric gas into space; and intense radiation baths at the planet's surface, among others. Each such event would greatly set back, if not stop, the evolution of life on an Earthlike planet, making these worlds unlikely habitats for humanity.

HOME IN THE GALAXY'S BAR

Moving a safer distance from the more dire effects of the supermassive black hole in the Galaxy's nucleus, we next encounter the bar of stars, gas, and dust

---

* There is no direct evidence that such a close passage has induced any of the mass extinctions that have occurred on Earth, but such passages are possible and have been hypothesized as the origins of some of these events.

† The age of our galaxy is not well known. It formed in parts, with the nucleus first, perhaps thirteen billion years ago, followed by the disk containing the spiral arms between seven and ten billion years ago.

(Figure 7.1). The existence of this bar, suspected since the 1990s, was confirmed in 2005. Its dimensions are roughly 24,000 light-years by 6,000 light-years. The matter in the inner region of the bar flows perpendicular to it, whereas in the outer region of the bar, stars, gas, and dust flow down the bar, turn around at the end, and flow back. Like a propeller, the bar itself also rotates about once every 15 million years. Some of the gas and dust in the bar are apparently channeled down to the supermassive black hole, providing raw material for star formation, as just discussed.

Several features of the bar suggest that few life-supporting planets could exist in it. Consider a star and planets that just formed in the bar and are moving down it from one tip to the other. As this system nears the central region, it is likely to encounter stars that are moving across the bar, perpendicular to its motion. As discussed earlier, sufficiently close encounters will change the orbits of planets, taking the potential Earth out of the Habitable Zone, or at least throw lots of comets toward the planets. These things are likely to occur eventually, even if this doesn't happen on the first hundred passes through the central region of the bar, long before advanced life evolves on the planet. Likewise, things get very hectic at the ends of the bar, where stars and clouds are turning around.

A related problem in this inner region of the galaxy is the large quantities of interstellar gas and dust in the bar that are being drawn down toward the black hole. We know this material is there because we have observed young stars near the black hole, implying that they formed from interstellar matter in the region. Among the matter that is found in interstellar space there is a cloud of antimatter, discovered in the 1970s. This is gas composed of particles that are identical to normal protons and electrons, but with the opposite charges. The details of how these antiparticles are formed are still unclear, but they apparently have to do with binary star systems (two stars orbiting each other) in which one of the pair is a black hole or compact stellar remnant called a neutron star.

The antimatter in the inner region of our galaxy is continually colliding with normal matter. For example, electrons hit their counterparts called positrons, and protons hit antiprotons. When such collisions occur the two particles annihilate each other and emit high-energy radiation in the form of gamma rays. The inner region of the galaxy is therefore bathed in lethal gamma rays. There is a variety of other mechanisms for creating a significant flow of intrinsically dangerous gamma ray, x-ray, and ultraviolet radiation in the galaxy's nucleus. As one example, the gases being pulled inward by the black hole are compressed, heated, and as a result, give off a lot of radiation. All this radiation is harmful to life that encounters it.

Let's next consider the possibility of life-sustaining planets forming around stars that are above or beyond the galaxy's spiral arms.

HOME ABOVE OR BEYOND THE SPIRAL ARMS

Despite appearances (e.g., Figure 7.1), galaxies like the Milky Way are much more than just disks of stars, gas, and dust with spiral arms. There are things in the galaxy above and below the disk and things that exist far beyond the visible limit to the disk. Most notable among the galaxy's constituents outside the disk are the globular clusters introduced earlier in this chapter. Containing up to hundreds of thousands of stars that are gravitationally bound together, globular clusters swarm around the disk like mosquitoes in summer.

Globular clusters regularly pass through the disk, but they spend most of the time outside of it in what is called the halo of the galaxy. This is a spherical region centered on the supermassive black hole in the nucleus of the galaxy, extending at least twice as far out in all directions as do the spiral arms. The halo also has billions of isolated stars that are so dim compared to the light from globular clusters that we do not normally see them.

The problem in creating an Earthlike planet outside the disk in which we exist today is finding enough gas and dust concentrated in one place out there to create the Sun and planets. Terrestrial planets form almost entirely from the debris of stellar explosions, as we discussed in Chapter 5. We know this because the matter in the universe when it first formed was only hydrogen, helium, and lithium. There is very little of any of these elements on Earth compared to iron, silicon, and numerous other elements, all of which are formed by fusion in stars and stellar explosions. A large fraction of the matter ejected when stars explode in the disk of the galaxy remains in the disk, allowing this gas and dust to be recycled into new stars as it passes through the spiral arms. Our solar system must be at least a second-generation system, because so much of the Earth's content must have been formed in a previous generation of stars.

The thing that prevents Earthlike planets from forming in the halo is that after the first generation of stars formed out there when the universe was young, the gas and dust ejected when these stars exploded did not remain sufficiently concentrated so that another generation of stars could form from it. Even in globular clusters, the gas ejected by exploding stars drifted away. Astronomers therefore see very few, if any, new stars forming outside the realm of the spiral arms.

HOME ELSEWHERE IN THE SPIRAL ARMS

I hope I have made the case that it is exceedingly unlikely a planet will maintain a suitable habitat for life over the billions of years necessary for sentient life to evolve on it in either the nuclear region of our galaxy or outside the region spanned by its spiral arms. However, astronomers have reason to believe that there are other places in the spiral arms in addition to where we are that are suitable for the formation of life-supporting planets. Apart from having only infrequent supernova explosions and infrequent near-misses between stars, these are regions rich in metals* created in earlier generations of stars.

The volume of the galaxy suitable for the formation and evolution of life constitutes the Galactic Habitable Zone. Because our solar system is located roughly in the middle of the realm of the spiral arms, I am going to create worlds near what I believe are the Galactic Habitable Zone's two extremes (outer and inner edges). We explore planets in both regions to see how they differ from Earth.

Synthesizing the discussion above and from previous chapters, there are four issues that we need to consider in justifying the existence of a life-supporting world elsewhere in the spiral arms of our galaxy, namely: are there enough and suitable metals to create terrestrial planets and life, are the

---

* Here meaning elements other than hydrogen and helium.

interstellar gas and dust sufficiently concentrated so as to condense into new generations of stars, is there a mechanism to initiate the star-formation process, and, is it likely that the life-sustaining planet will avoid life-destroying damage and disturbance for the four and a half billion years necessary for sentient life to form? Examples of life-destroying damage include collisions with large asteroids or comets billions of years after the planet forms and supernova explosions within a few light-years of the planet after it has advanced life on its surface. Examples of life-destroying disturbances are gravitational tugs that change the planet's orbit so that the temperature ceases to allow liquid water on its surface.

Life on the Outer Edge—Kern I

We don't know the details of how our galaxy was assembled. This knowledge would help clarify how different the outer and inner regions of the galaxy's disk are from our neighborhood in space. The best we can do now is to observe distant galaxies that are in the process of forming and use their properties to deduce how the Milky Way came together. We can make such observations because the farther away we observe things, the farther back in time we see them.

Young galaxies seen at the farthest reaches that we can presently observe strongly suggest that galaxies like the Milky Way form from the inside out. If this is correct, it implies that the galaxy built up from the merger of smaller ensembles of stars, gas, and dust. This implies that the inner region of the spiral arms developed earlier than the outer regions. Therefore, more cycles of star formation and destruction have occurred in the inner region, resulting in the inner region having more metals than the outer edge of the arms. It is in the outer region that we first explore an alternative Earth.

The Earthlike planet Kern I formed near the outer edge of the spiral arms. Gases orbiting in the outer reaches of the arms are moving more slowly than the gases that formed the solar system. Therefore, the gases in the outer regions pass less frequently through the spiral arms. Because this passage is what leads to star formation, star formation occurs more slowly out there than here. It took extra time for cycles of star formation in the outer part of the galaxy to generate the metals needed for Earthlike planets compared to stellar evolution in our neighborhood. Therefore, Kern I formed two billion years after the Earth.

Kern I is located 50,000 light-years from the nucleus of the galaxy (see Figure 7.1), which puts it 24,000 light-years beyond the orbit of the solar system around the galaxy's center. Kern I is surrounded by fewer stars than we are on Earth. As a result, it is less likely that Kern I would be near enough to stars that explode as supernovae for these explosions to damage the planet in ways that could lead to mass extinctions of life than occurred here. There is evidence that at least one of the mass extinctions on Earth, the Ordovician–Silurian extinction 450 million years ago, may have been caused by an especially powerful supernova explosion called a gamma ray burst; however, I posit that no mass extinctions on Kern I occur because of a supernova. Mass extinctions are still likely to occur there for other reasons, such as impacts of large space debris and temperature and atmospheric changes caused by activity inside Kern I.

The lower density of stars in the vicinity of Kern I also justifies the assumption that fewer stars pass close enough to Kern I's Oort comet cloud than has occurred for our solar system. These events cause many pieces of space debris to hurtle inward, therefore they increase the probability of impacts. With fewer of them, impact-driven mass extinctions are less likely to occur on Kern I than did here.

I therefore assert that Kern I has two fewer mass extinctions than did the Earth. As a result, sentient life formed about 200 million years earlier in the life of Kern I than it did in the life of the Earth. Paleontologists find that each mass extinction leads to the destruction of numerous species* and the evolution and blossoming of others. To connect Kern I to the Earth as closely as possible, let's assume that the evolution of life on the two planets is initially identical.† Upon occurrence of the first mass extinction on Earth that does not occur on Kern I, the evolution of life on the two worlds diverged irrevocably. Of course, that and the second extra mass extinction were essential in allowing homo sapiens sapiens (namely, us) to evolve into existence on Earth.

---

* There is a hierarchy in the classification of life-forms, including species, genus, family, order, class, phylum, kingdom, and domain. We need not explore this scheme for our purposes.
† Strictly to understand the consequences of fewer mass extinctions, not to suggest that it could really happen.

Therefore, the sentient life on Kern I will have a profoundly different history than we did. It is unlikely in the extreme, for example, that they will have the same mammalian ancestors that we have.

What will sentient Kern I'ians see of the Milky Way Galaxy when they become aware of it nearly two billion years from now? First of all, they will see only a few of the stars visible to the naked eye from Earth. The stars that we see today are all within roughly 15,000 light-years of our planet. With Kern I always at least 24,000 light-years from us, only the stars that we can see which are located in the direction of the outer reaches of the galaxy could be visible to them without the aid of a telescope. (They can see stars in our direction that are within 15,000 light-years of them. When we are closest to them, these same stars are only 9,000 light-years away from us, so we can both see them.) Whenever Kern I is not directly outward from us in the galaxy (e.g., when it is on the opposite side of the galaxy), they will be too far away from the solar system to see any of the stars in our night sky.

Furthermore, some of the stars in our sky will have exploded by the time sentient creatures live on Kern I. At most, Kern I'ian astronomers will see remnants of these stars. Therefore, Kern I'ians will see a very different set of stars in their night sky and, hence, different patterns of stars than we do. It is likely that as with humans, their brains will use pattern recognition to help them survive and, therefore, they are as likely as we are to see patterns even when they don't exist. So Kern I'ians will plausibly identify asterisms in the night sky.*

The Milky Way will also look different to them. People who are lucky enough to see the night sky in especially dark areas on Earth see the glow of stars and gas in the Milky Way spanning the heavens. We see it because it is the plane in which the disk of stars, gas, and dust revolve, along with the spiral arms that have been induced in it (as discussed in Chapter 5).

Much of the "Milky Way" in our night sky is glowing clouds of gas and dust in nearby spiral arms that are lit by bright stars near them. We are between two spiral arms: the Sagittarius Arm on the side of the galaxy closer to the center and the Perseus Arm on the outside of the galaxy from our perspective (Figure 7.1). Because there are spiral arms filled with glowing gas on both sides of us, we see the glow of the galaxy as a ring of light that

---

* Asterisms are the patterns we see, such as Orion or the Big Dipper.

surrounds the Earth. Put another way, because different parts of the sky are visible during different times of the year, no matter when we look, we can see light from the spiral arms. The same is not true on Kern I.

People on Kern I will see only half the sky emblazoned with light from the Milky Way. Their world is an outpost created on an outer edge of a spiral arm near its tip. When they are looking toward the center of the galaxy, Kern I'ians will see the same kind of glow we see in the sky, because they will be looking into a spiral arm. However, when they are looking away from the center, they will be viewing only the outer region of the galaxy and beyond, in which there are not enough stars or interstellar gas and dust to interact and thereby glow.

Until the early twentieth century, humans took the perception of being surrounded by the glow of the Milky Way as an indicator that Earth is in the center of our galaxy (which in those days was also considered to be the entire universe). Kern I'ians will clearly see that things are different in space on different sides of the celestial sphere. With one side of the sky filled with a band of light and myriad stars and the other side dark and with fewer stars, clearly Kern I is on the edge of something. That perception would surely lead to another whole mythology, perhaps that Kern I is on the boundary between two different parts of the universe. The side with glow from the Milky Way might be deemed to be a region where too much activity was occurring for the good of their race, and the other side is a region with too little activity.

Because the sentient beings on Kern I arrive nearly two billion years from now, they will not see some things that we see today. Some major differences stem from the fact that our galaxy is a cannibal. The Milky Way is sufficiently massive so that smaller galaxies passing in the night are eaten by it. It works like this: as a smaller galaxy flies through our galaxy, the gravitational attraction of the stars, gas, dust, and other components of it pull the victim galaxy apart. One effect that causes this destruction is called dynamical friction, which basically means that as the smaller galaxy passes through the larger one, the smaller galaxy forces some of the bigger galaxy's stars to move and collect behind the passing smaller one. This trail of stars behind the moving smaller galaxy creates an extra gravitational tug on this victim galaxy, slowing it down (Figure 7.3a,b).

Even if the smaller galaxy passes through the plane of the spiral arms

(a)

Intruder

without being completely destroyed (and they usually do pass through when they first encounter our bigger galaxy), some stars are stripped off the smaller one, either remaining in the Milky Way or trailing behind the little galaxy of their origin. Furthermore, once the smaller galaxy is slowed, it is very likely to orbit around our galaxy and then pass through it again. This process goes on until the smaller galaxy loses sufficient kinetic energy so that it becomes part of the bigger galaxy. In this way our galaxy is today cannibalizing several smaller galaxies.

It is likely that by the time sentient life exists on Kern I all the galaxies we are eating will be consumed and digested. Therefore, either Kern I'ians will see a more pristine Milky Way, without the loops and tendrils of stars above and below it that we see today (Figure 7.4), or they may see the Milky Way consuming other galaxies.

Speaking of galaxies, we noted in Chapter 5 that galaxies in clusters are gravitationally bound together, meaning that they orbit each other. Nearly two billion years from now the galaxies in the cluster in which the Milky Way resides, called the Local Group, will have moved enough so that their locations will be significantly different from what they are now. For example, consider the Triangulum Galaxy, a spectacular normal (unbarred) spiral in the Local Cluster that is barely visible to the naked eye in the northern hemisphere. Kern I'ians will see Triangulum clearly as a distinct fuzzy blob in their sky. They will also see the Andromeda galaxy much

This clump of stars pulls on the
intruder, slowing it down

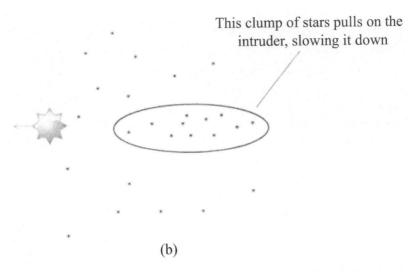

(b)

Figure 7.3a,b: Dynamical friction is caused when a massive body, such as a galaxy or a particularly massive star, passes lighter ones, pulling the latter into new positions, from which they pull back and slow the massive body down.

larger and brighter than it is today. I have more to say about this last point in Chapter 10.

People on Kern I will also see that some of the galaxies in the Local Group have different shapes than we see today. As we discussed earlier, some galaxies are elliptical, meaning that if you could see each one from a distance, it would look more or less egg-shaped. If you look down the long axis of an egg, it looks circular, whereas from the side, it looks oval. The same is true with elliptical galaxies. In nearly two billion years what we see as a circular galaxy (an elliptical seen from one end of the "egg") will look oval to Kern I'ians. There are also other galaxies, the irregulars such as the Large and Small Magellanic Clouds, that will look completely different to Kern I'ians (although we can't predict exactly what they will look like).

Let's now turn to another alternative world in our galaxy, this one forming close to the bar.

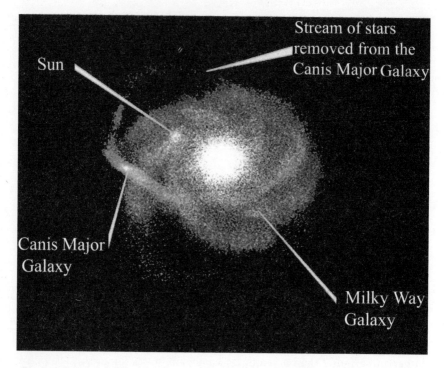

Figure 7.4: Debris left by the Canis Major Galaxy as the Milky Way cannibalizes the smaller galaxy. COURTESY: R. IBATA/STRASBOURG OBSERVATORY, ULP/ET AL., 2MASS, NASA.

Life on the Inner Edge—Kern II

Kern II and the star system in which it lives* were born in a spiral arm of the Milky Way just outside the bar of stars, gas, and dust that goes through the center of the galaxy (Figure 7.1). Because that region of the galaxy apparently formed earlier than the region in which we reside, the metals necessary to form terrestrial planets and complex life were available earlier there than they were here. Hence Kern II came into existence a billion years before our solar system did. As we now show, this does not necessarily imply that sentient life evolved there a billion years before we did.

Kern II is located about 14,000 light-years from the nucleus of the Milky Way and therefore the closest that we come to it, when we are on the same side of the galactic nucleus as it is, is about 12,000 light-years. The vast ma-

---

* Hereafter I refer to both the planet and this system as Kern II.

jority of the stars, gas, and dust in the realm of Kern II have circular orbits around the nucleus of the galaxy, unlike the motion of the stars and other material in the bar itself, which ends some 2,000 light-years from that world. Recall that in the bar, things flow mostly toward or away from the nucleus.

Because Kern II is closer to the center than we are, a simple calculation based on Newton's law of gravity reveals that Kern II orbits the center of the galaxy much more quickly than we do. Our solar system orbits it about once every quarter of a billion years. Kern II's system goes around once every 21 million years.

Whereas Kern I is in a more sedate environment than Earth, Kern II faces more external dangers from radiation and impacts as a result of its neighbors than we do. Orbiting more rapidly than Earth, Kern II goes through the galaxy's spiral arms more frequently than we do. Most stars form in spiral arms, so as Kern II enters an arm, new stars will be forming near it. Among these will be some stars with between roughly 9 and 100 times the mass of our Sun. These stars end their relatively brief lives as supernovae. Because Kern II won't have enough time to leave a spiral arm before these high-mass stars end their lives, that world will be exposed to the radiation and blast waves of more supernovae than either Earth or Kern I experience. (It isn't that we avoid such events, but rather that we encounter them less often inasmuch as we spend less time in spiral arms.)

Unless Kern II is exceptionally unfortunate, the supernovae near it will occur far enough away so that the planet isn't physically destroyed (unlike the supernova that destroyed Romulus in the movie *Star Trek*), but they will cause more mass extinctions and hence will delay the evolution of sentient life there. Whereas Kern I experienced no mass extinctions due to supernova, I set Kern II to encounter two more than we did. Because the damage from these events can apparently be wide-ranging (including destruction of the ozone layer, direct high-energy radiation to the planet's surface, and heating and removal of some of the atmosphere, among other things) I delay the evolution of sentient life there by a billion years compared to what would have happened without the extra supernovae.

Because Kern II was formed a billion years before Earth, the extra extinctions on Kern II that change and delay evolution there led to the formation of sentient life on both Earth and Kern II at about the same time. Furthermore, I posit that the Kern II'ians have intellectual ability similar to ours, so that they, too, can understand the workings of the cosmos. It is nevertheless likely that the astronomical environment of Kern II will delay at least some of their scientific understanding compared to ours.

To understand the differences in the astronomical environment that Kern II experiences when sentient life evolves there compared to ours, we need to set the astronomical stage. There will be periods of millions of years when Kern II is passing through one spiral arm of the galaxy or another. During such times it will be surrounded by more bright stars than we ever are. The brightest stars around us are all at least 30,000 times dimmer in our sky than our Moon. The light from a dozen hot, young, bright stars as close as the closest star in our sky would provide Kern II with so much light that even when its star and moon are down, Kern II will still have a blue sky, albeit darker than ours. It is during such a period that sentient life evolves onto Kern.

What might Kern II'ians think about the universe if their sky never turned black and so they never saw a star other than their own and the dozen others that light up the night sky? Indeed, they would also not see the other planets, comets, galaxies, or the band of light we call the Milky Way. The mystery of what lies behind the blue veil overhead would deepen when they saw the occasional especially bright shooting star or, as happened here in 1054 AD, the occasional extra light in the sky created by supernovae that would appear and then fade over a matter of weeks.

Shooting stars and visible supernovae are likely to lead Kern II'ians to conclude that there is more "out there" than meets the eye. But what? Without seeing other planets orbiting their star, it would be impossible for Kern II's Kepler to work out the laws guiding orbits. Newton, building on Kepler's work, would have had a harder time working out the law of gravitation and he would have been less motivated to improve the telescope, which was invented on Earth as a military spying device, but was quickly turned upward to explore the cosmos. Without seeing thousands of stars, astronomers of Kern II's twentieth century would be unable to find the classification scheme of stars that has led astronomers on Earth to develop scientific explanations of why stars shine and how they evolve. Without seeing other planets in its star system, astronomers would be unable to test a theory of how Kern II, its moon, and its star formed and how they interact.

These delays in scientific understanding of the cosmos would have in turn delayed the development of many areas of science and related technologies. Not seeing the sky could well delay their understanding, indeed their knowledge, of the existence of the universe until they venture into space. This raises the question of why they might develop rockets capable of going into space at all (as opposed to low-flying rockets for military use, as were the first rockets used on Earth). I can envision two scenarios.

First, pilots in very high-flying aircraft might glimpse bright planets and stars that can't be seen from the ground on Kern II, as also occurs here on Earth. Knowing that there is something else "out there" would then stimulate the effort to learn more about it, which would lead to the development of rockets. Second, emerging technological civilizations on Kern II would have need for satellites in orbit for communication. In any event, it is only when they get above the atmosphere that they will become aware of all the things astronomical that we have seen as long as there have been people on Earth.

Once their eyes are open to the universe, Kern II'ians will move forward in their science and eventually they, like us, will begin searching for other life in the universe. First they will observe stars wobbling due to the gravitational pull of orbiting planets. Then they will measure the Doppler shift (see Chapter 5) of the stars that occurs as the planets orbit them. Then they will see individual giant planets, such as Jupiter, orbiting other worlds. Then they will see terrestrial planets, including Earth.

Imagine their seeing our planet right now. They would observe that it is at the right distance from the Sun for water to be liquid here. The radiation emitted from the Earth and scattered sunlight from its atmosphere would tell the Kern II'ians that this planet is similar in physical and chemical properties to their own world. Plausibly, they would reason, Earth could support life. Encouraged by this discovery, they would launch gigantic telescopes into space able to see details on Earth. Recall that Kern II is always at least 12,000 light-years away from us (farther when they are on the opposite side of the galactic nucleus from us). Because light, radio waves, and all other electromagnetic radiation emitted from the Earth travel at a finite speed, they would be seeing the Earth as it was 12,000 years ago. What would they see on the Earth back then?

The last great ice age was in the process of winding down, so they would see more ice on the Earth than there is today. They would see giant woolly mammoths roaming North America and Stone Age people and other species of the *homo* race scattered over the planet. They would see some animals being herded. They would see villages and nomads. They would see forest fires that raged for weeks and even months. They would see lightning flashing and volcanoes erupting. They would see a world in progress.

8   What If the Sun Were Less Massive?

HIGHLIGHTS FROM THE STORY OF THE HAANSEN MISSION TO THE DARKSIDE

For eight months the grinders had been carving a hole through the ice mountain that separates the half of the world in which life was known to exist from the half that no one had ever visited, or at least returned to tell about. The Committee on Science and Technology had decided three years ago that now was the time to find out what life, if any, existed there. It had taken a year to raise the funds and longer still to design and build the drilling equipment, as well as the life-support equipment that was thought to be needed for the explorers.

The wood-fired, steam-driven augers had done their job well, helped in part by the heat from the engines, which had been used to smooth out the ice tunnel and to provide drinking water for the construction crew. The fire in the firebox had been their only source of light and heat in the pitch-black bowels of Mt. Cameron.

They broke through to the other side of the mountain without any warning. The drillers had been instructed to return to base immediately when that happened, or to at least send a telegraph message via the cable they laid down as they went. Needless to say, they did neither. Backing the auger into the tunnel, the four men walked to the opening they had created and looked, or tried to, onto a world lit only the slightest amount by light diffusing through the clouds. Moonlight would have helped, but as luck had it, the moon was over the side of the planet lit by their star, Bantam. Frustrated, they sat down and had lunch.

Word of the breakthrough reached the station that had been built at the opening of the hole an hour later. Within half an hour, reporters and photographers were swarming over the site, waiting the arrival of the six explorers, led by Liam Haansen, and their dog teams. With auger machines tucked to the side of the tunnel, Liam and his team stood in a line for a round of photographs, then they waved farewell and entered the tunnel, now illuminated by a feeble electric light worn by Liam, who led the way.

They emerged on the Darkside sixteen hours later, exhausted and chilled to the bone. Liam decided that they would set up camp in the tunnel, rather than test the stability of the mountainside while they were so tired. Feeding the dogs, the explorers pitched tents to cut the wind that howled through the tunnel. As expected, their biological clocks failed to awaken the explorers in the dark world that they had entered. A mechanical clock managed to ring its alarm, but so deep was the cold that something inside became brittle and snapped.

The team started down the short distance to the bottom of Mt. Cameron, moving diagonally across the slope of the mountain so the dogs and their sleds would not tumble. Fortunately, it had snowed recently, making the descent relatively easy and uneventful. They began sledding away from the mountain, through a valley separating two snow-capped hills. For fifteen days, they worked their way eastward, away from light, heat, and civilization. During all that time, they were sliding over icy hills and valleys, with the light visible through the omnipresent clouds growing dimmer as they moved farther from the Bantam-lit side of their world.

On the sixteenth day, the clouds began to thin and they became the first people of their time ever to see a star-filled night sky. They had, of course, seen a few bright stars during the darkest hours of the year, but nothing like this. So awe-inspiring and humbling was the experience that the scribe, Bert Rifkin, the most sensitive among the travelers, began to cry, which he much regretted as the tears froze on his beard.

Later that day, the moon began rising, which helped their view of the night-side of their world tremendously. They had by this time left the mountains behind them and were working their way over a snow-strewn plain. What they saw was not encouraging. They were in a field of ice spires, like branchless trees. Everything was gray, except for the occasional turquoise glow caused when moonlight struck the ice at the right angle. It was there that they saw on the eastern horizon a faint red glow.

Kris Vallejo, navigator for the expedition, triangulated on the most distant point in that direction, reporting that it would be at least a two-day trip, taking them to the limit of travel, considering the supplies they had with them. Liam decided that they would make the red glow their final destination before returning.

The next day the snow-covered ice gave way to rocky terrain strewn with boulders, some twice as high as a man. They fixed wheels to the sleds and proceeded. On the final day outbound they reached a steep conelike wall, above which the glow was bright and the air was comfortably warm. They walked around the cone, looking for a way up or through it. The journey around revealed that the cone was six miles across and impossible to mount. Choosing a hill overlooking it, the team began climbing. Halfway up the hillside they were able to see over the top of the cone into a bubbling sea of red-hot lava rising from the center of an enormous crater. The lava flowed toward the cone wall and then apparently descended back into the planet. "It is a convective lava tube," explained team geologist Faniell Jones.

Upon reaching the top of the hill they made the discovery that changed everything. It was a tiny village, with three round-topped build-ings, pens for animals, and the remains of sixteen men, women, and children, and nearly sixty dogs.

As the powerhouse that ignited the formation and evolution of life, the Sun has done a pretty good job. Not that it is perfect, mind you, as solar flares, coronal mass ejections, and possibly Sun-induced ice ages reveal, but all in all, it has done well by us. Could stars with different physical properties have done as well in enabling life to form and then in sustaining it?

The essential connection between the Sun and life on Earth is the energy our star provides us in the form of infrared radiation, visible light, and ultraviolet radiation. In everyday terms, we measure the Sun's energy in the same way as we would the energy of a fire: from its area and surface temperature. These two parameters combine to determine how much energy it emits into space. The Earth, as we discussed in Chapter 2, is located in the Sun's Habitable Zone, where the solar energy enables water to be liquid on our planet's surface.

Including ultraviolet radiation as "essential" in the paragraph above is important, but complicated. To give us a peek at what lies ahead and to address the obvious concern that we are taught that ultraviolet radiation is to be avoided, let's briefly consider the effects of this type of radiation. The young Earth's atmosphere had virtually no oxygen and, hence, no ozone to prevent ultraviolet radiation from penetrating to the lower layers of the air and to the Earth's surface. Ultraviolet radiation undoubtedly both aided and hurt the formation of life here. Those photons have enough energy to enable a wide variety of atoms and molecules to combine in the young Earth's atmosphere and in the upper layers of its oceans. Nearly as effectively, ultraviolet photons tore molecules apart. However, it and other energy sources, notably lightning, powered the formation of life.

Eventually plant life formed and gave off oxygen as a waste product. The early oxygen combined with minerals in the water and on the surface of the planet, but eventually large amounts of it settled in the atmosphere in the form of O_2 oxygen molecules. Ultraviolet radiation from the Sun split some of these molecules. The resulting individual oxygen atoms then combined with other oxygen molecules to form ozone, O_3. Ozone is especially good at absorbing ultraviolet radiation, greatly reducing the amount of it that reaches the surface today compared to when there was no ozone.

With that background in hand, let's consider the possibility of creating a life-bearing world orbiting a star emitting different amounts of infrared, visible, and ultraviolet radiation than does our Sun. Is this even possible? Are there stars with different intensities of electromagnetic radiation than the Sun? A quick look at stars with different brightnesses in the night sky suggests that different stars do, indeed, emit different amounts of light, but that could be deceiving. The stars we see are at different distances from us. Perhaps they all intrinsically emit the same amounts of all kinds of radiation, with the closer ones just appearing brighter and the farther ones appearing dimmer. An evening with a clear sky and a strong pair of binoculars or a telescope will show that despite being at different distances, we can be sure that stars do emit different amounts of electromagnetic radiation. The clue is in their colors.

The colors of stars and other hot objects are determined by their temperatures. This is easy to visualize when you turn on an electric stove. As the heating element warms up, the first color it glows is red. As it gets hotter, it becomes orange and then yellow. The coils on a stove are heated on their surfaces by the flow of electricity through them. Stars, on the other hand, are heated by nuclear fusion deep inside them, where temperatures exceed 10 million K (18 million°F). Why then are they at millions of degrees inside, but cooler than 12,000 K (21,000°F) on their surfaces?

The answer lies in the energy lost as the hot gases transfer their heat outward. The heat deep inside the stars comes in part from the gases that comprise the star compressing its inner regions. The more you compress any gas, the hotter it becomes. The rest of the heat comes from energy released in the form of gamma rays when particles undergo nuclear fusion and thereby come together to form heavier particles. The Sun, for example, fuses hydrogen into helium. The newly formed gamma rays slam into nearby gas particles and are absorbed by them, pushing the particles outward. This has the effect of thereby preventing the star from shrinking. Without that activity, stars like the Sun would be much smaller than they are. The rapidly moving gas particles in turn hit other gas particles. This interaction causes them to emit photons. Because the gas particles lose energy as they prevent the star from collapsing, these photons are less energetic than the gamma rays created by fusion. The cycle of photon absorption and then photon emission by particles repeats billions of times, eventually bringing photons toward the stars' surfaces.

Recall from Chapter 5 that the energy of a photon is measured by its wavelength: the higher the energy, the shorter the wavelength. The highest-energy photons are gamma rays, followed by x-rays, ultraviolet, visible light, infrared, microwaves, and radio waves. By the time photons reach the star's photosphere, most of them are either infrared, visible, or ultraviolet, rather than gamma rays.

Like the Sun, most stars are fusing hydrogen into helium in their cores. Called Main Sequence stars, they have the property that the hotter their cores, the hotter their surfaces. Furthermore, their core temperatures depend on their masses: the higher the mass, the higher the core temperature. In summary, the highest-mass Main Sequence stars have the hottest cores and surfaces, while the lowest-mass Main Sequence stars have the coolest cores and surfaces.

Astronomers discovered that each star has a wavelength at which it emits light most intensely (Figure 8.1). The wavelength of the peak can be in any part of the spectrum, depending on the star's surface temperature. The hotter a star, the shorter the wavelength of the peak of its radiation intensity compared to the wavelength of the peak of a cooler star. The wavelength of the peak of intensity for stars of different temperatures was first worked out in 1893 by German physicist Wilhelm Carl Werner Otto Fritz Franz Wien. The coolest stars have peaks in the infrared; stars with intermediate temperatures peak in the visible part of the spectrum; and the hot stars peak in the ultraviolet.

Another relevant feature of their radiation is that all stars emit photons at all wavelengths. Therefore, stars that peak in the infrared or ultraviolet are still visible to our eyes. The coolest stars, with peaks in the infrared, emit most intensely in the red (Figure 8.1), whereas a sequence of progressively hotter stars that peak in the visible have red, orange, yellow, green, blue,* or violet peaks. Stars that peak in the ultraviolet emit violet light most intensely among the visible colors.

Furthermore, the hotter a star, the more photons of every wavelength it emits, compared to cooler stars. This was first explained theoretically by German physicist Max Planck. The mathematical formalism that describes all of this is called the theory of blackbody radiation.† In summary, the surface temperature, the wavelength of the most intense emission of photons,

---

* Astronomers don't use indigo as an individual color.
† This name has nothing to do with race. A blackbody absorbs all the radiation that strikes it, thereby appearing black, which is the absence of color. The name was coined by Gustav Robert Kirchoff (1824–1887), who did some of the pioneering work on the subject.

The higher the temperature of a blackbody, the shorter the wavelength of its maximum emission (the wavelength at which the curve peaks).

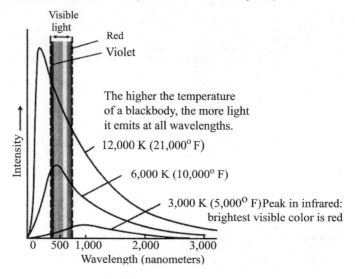

Figure 8.1: Intensity curves for blackbodies of different temperatures. Visible colors are: red, orange, yellow, green, blue, and violet. CREDIT: W.H. FREEMAN & CO.

and the total number of photons of all wavelengths (i.e., the brightness) are all determined by a Main Sequence star's mass. The mass also determines the total amount of energy generated inside it by nuclear fusion (introduced in Chapter 5).

Returning now to the night sky, we see stars with different colors because they have different surface temperatures. Therefore, we have a priori evidence that not all stars are equal. But stars are at different distances from us. The colors are different, but we still need to know the total amount of energy they emit, which also depends on their sizes, as noted earlier. To get this information, we need to visualize them as they would appear if they were all side by side.

We can't move the stars, of course, but if we know how far they are from us, we can measure their brightness as seen from Earth and then use a little algebra to calculate how bright they would be if they were all the same distance away from us. The ability to determine distances (introduced in Chapter 2) is done most reliably in astronomy by using two telescopes in different

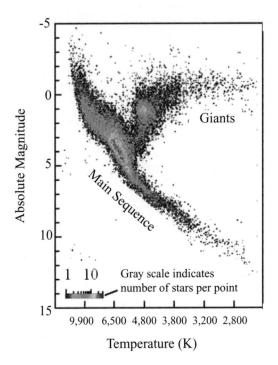

Figure 8.2: Hertzsprung–Russell diagram showing the relationship between stellar brightnesses (absolute magnitudes) and surface temperatures. COURTESY: EUROPEAN SPACE AGENCY.

places. They see the star at different angles (see Figure 2.5). This parallax in turn yields the star's distance.

The brightnesses that stars have as seen from Earth, regardless of their distances from us, are called *apparent magnitudes.* The brightnesses they would have if they were all put side by side are called *absolute magnitudes.* This latter concept is physically meaningful because it tells us how much energy stars are emitting relative to each other. At the beginning of the twentieth century the concept of absolute magnitude was proposed, and by 1913, it had been applied to a diagram that provides the key to understanding which star we should choose for this chapter.

The Hertzsprung–Russell diagram, discovered independently by Ejnar Hertzsprung (1873–1967) and Henry Norris Russell (1877–1957) plots the stars' absolute magnitudes on the vertical axis and their surface temperatures on the horizontal axis (Figure 8.2). Because stars are not randomly distributed on this diagram, it shows us that these two parameters are related. The two regions of the figure we need to worry about in finding other stars around which life-supporting planets might orbit are labeled "Main Sequence" and

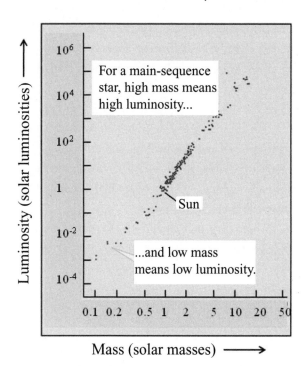

Figure 8.3: *Mass–luminosity relationship for Main Sequence stars.*
CREDIT: W.H. FREEMAN & CO.

"Giant." About ninety percent of all stars lie on the Main Sequence, and most of the rest are Giants.

One last observational property of stars is essential for us. Astronomers have been able to determine the masses of stars by observing their motion in binary star systems. Binaries are pairs of stars that were formed together and that orbit each other. Using equations derived by Newton, careful observations of the periods and separation of certain binary systems provide the masses of the two stars in them. From this information, astronomers discovered that for Main Sequence stars (and Main Sequence stars, only) there is a direct relationship between the star's mass and its absolute magnitude (or, equivalently, its luminosity, which is the total energy it emits per second). This is called the mass–luminosity relation, which states that the higher the mass of a Main Sequence star, the greater its luminosity (Figure 8.3).

Based on observations of the cosmos, along with theoretical predictions in the burgeoning field of nuclear physics, and data presented in the Hertzsprung–Russell diagram and the mass–luminosity relation, astrophysicists in the early twentieth century developed theories of why stars shine and how they evolve. Although this work continues today, the basic concepts we need are now well understood. Here are some of the relevant results that have been confirmed innumerable times in the past century.

- The chemical composition (by mass) of interstellar clouds is typically 74.5 percent hydrogen, 24 percent helium, and traces of all the other naturally forming elements. Put another way, about 92 percent of the particles throughout our galaxy are hydrogen, and 7.8 percent are helium, with traces of the others.
- The total mass of the gas and dust that comprise a star determines the temperature everywhere inside it. The heating is caused by the gravitational attraction of the star's mass compressing it.
- Wherever the temperature is greater than roughly 10 million K (18 million°F), four hydrogen atoms will fuse together to create helium. Energy in the form of gamma ray photons is emitted in this process.
- The temperature exceeds the critical 10 million K only in the central regions of Main Sequence stars, called their cores.
- Main Sequence stars are those that are in the first stage of stellar evolution, having formed from small fragments of interstellar clouds.
- The lowest mass that a Main Sequence star can have is 0.08 solar masses. A blob of gas and dust below this amount of mass cannot compress its core to the 10 million K required to initiate fusion. Objects with less than 0.08 solar masses do exist, by the way, but they don't shine due to fusion. You can envision them as overgrown Jupiters.
- The more massive the star is, the larger the region of core fusion and hence the brighter the star (which explains the mass–luminosity relation).
- For stars with less than 0.4 times the mass of our Sun, the helium created by fusion bubbles out of the core and is replaced by hydrogen. This continues until the entire star is helium, after which fusion stops and the helium star, called a red dwarf, cools off forever.
- For stars with more than about 0.4 times the mass of our Sun, helium stays in the core. When all the core hydrogen is converted to helium, core fusion stops in Main Sequence stars.

- The cores of these latter stars contract, heating the hydrogen just outside the core until it begins fusion in a shell just outside the core. This is called "shell" fusion.
- Shell fusion provides energy in the form of photons that push the outer layers of the star out farther than they were on the Main Sequence, creating a giant star.
- When the Sun becomes a Giant, it will be so large in the sky that its heat will cause global warming that will make what is happening today seem like a walk in the park. The water in Earth's oceans will vaporize and life here will end.
- Main Sequence stars with less mass than the Sun (and more than 0.4 times the Sun's mass) expand into smaller-sized red giants than will the Sun.
- Although there is activity in stars through and after the Giant phase, life-sustaining planets around them will not be able to support life thereafter, so we need not worry about the later times.

I hope I have established that Main Sequence stars with different masses have different surface temperatures (as indicated by their colors). In this chapter I explore life on a planet around a star with less mass than the Sun.

There is one final issue: we need to know whether lower-mass Main Sequence stars will last long enough for sentient life to evolve on planets orbiting them. Intuitively, it seems that more massive stars are going to shine longer on the Main Sequence than lower-mass ones because the cores of more massive stars have more fuel. In fact, lower-mass stars last longer. The reason is that there is essentially a competition between the increased rate of fusion in higher-mass stars (due to their greater gravitational force heating their cores) and the amount of hydrogen fuel in the cores available to be fused in helium. Calculations reveal that higher-mass stars consume their core hydrogen so much faster than lower-mass stars that the lower-mass ones last longer. Because the Sun has shone long enough for us to evolve here, lower-mass stars will shine even longer, so it is physically plausible to consider an Earthlike planet orbiting one of them.

Using the Hertzsprung–Russell diagram and the mass–luminosity relation to determine the masses and surface temperatures of stars, we are now ready to explore whether a star with less mass than our Sun could support a life-bearing planet and, if so, how that world would be different than ours.*

The first order of business is to choose a mass for our new lower-mass star, named Bantam. We know from our survey of stars that we want a Main Sequence star, like the Sun. Whatever mass we choose for it, the Hertzsprung–Russell diagram reveals that Bantam will be dimmer and cooler than the Sun. To make matters interesting, I choose Bantam to have a mass one-quarter that of the Sun. Such a star will have a surface temperature of about 3,200 K, which is roughly half of the Sun's 5,800 K. It emits about one percent as much energy as does the Sun. Whereas the Sun's blackbody wavelength peaks in the blue-green part of the spectrum, Bantam's blackbody peaks in the infrared. As noted above, this does not mean that Bantam is invisible, but rather that its most intense visible radiation is red, so it will appear red in the sky of the planet, Dimune.

Bantam has about 0.36 times the diameter of our Sun. In other words, from one side of its photosphere to the other, Bantam measures about 311,000 miles. This is only 39 times the diameter of the Earth, compared to the Sun, which is 109 times the Earth's diameter.

The distance from Bantam that Dimune must be in order to have liquid water on its surface can be calculated based on Bantam's surface temperature, size, and the temperature we want for Dimune. The freezing point of water is 32°F (273 K) and its boiling point is 212°F (373 K). We want Dimune to experience temperatures similar to what we have on Earth, meaning that the average temperature on Dimune should be close to 62°F (290 K). Combining these yields an orbital distance of around fifteen million miles, compared to Earth's ninety-three million miles from the Sun. Put another way, Dimune is orbiting less than half the distance from Bantam that Mercury is from the Sun.

Based on exploring Mynoa's orbit around Tyran in Chapter 2, we need to check that Dimune is outside of the star's Roche limit. Recall that within this distance from Bantam, the tidal force on Dimune would be so great that the crust would be lifted right off the planet. This process of stripping the outer layers would continue until the entire planet was disassembled, creating a ring

---

* The converse question of whether the Sun could have more mass was explored in *What If the Moon Didn't Exist?*, also written by the author.

around Bantam. The Roche limit is about 250,000 miles, so Dimune orbits outside it, sixty times farther away.

Dimune is fated to have a much smaller moon than ours. The reason for this lies in the tides that the star raises on the planet. To begin, Dimune was formed as a body identical to the Earth. It was spinning at about the same rate as the Earth, about once every eight hours. The closeness of Dimune to Bantam leads to the question of how much higher the tides created by this star are on the planet compared to the height of the tides created by the Sun on Earth. The answer is that Bantam's tides on Dimune are sixty times higher than the tides created by the Sun on Earth.

This number applies for both the ocean tides and the land tides of solid rock (see Chapters 1 and 2). Before Dimune's surface solidifies, it will experience huge tides of molten rock. Indeed, we discussed this back in Chapter 2 in the context of Tyran creating such molten rock tides on the young Mynoa. When Dimune first formed in orbit around Bantam, the planet's surface was molten rock spinning very rapidly.

The tides of molten rock on young Dimune would be so high and the resulting friction between layers of rock so great that the slowing of Dimune's rotation would be in full swing before the collision between the planet and the asteroid that, in the case of the Earth, created our Moon. Instead of rotating once every eight hours when that impact occurred, Dimune would already have been slowed to a twelve-hour day. Therefore, the impacting body would not be able to splash nearly as much matter into orbit. Recall from Chapter 2 that the rapid rotation of the Earth is part of what enabled so much debris from the impact to form a ring around our planet, which quickly condensed into the Moon. Most of the debris splashed off the more slowly rotating Dimune will fall back onto it, similar to the events described in Chapter 3. To simplify what follows, I assume that the impact does not tilt Dimune, hence the planet does not have seasons.

The moon formed in orbit around Dimune by the impacting body is tiny compared to ours. With only $\frac{1}{100}$ the mass of our Moon, it is just over $\frac{1}{5}$ the diameter of that body. The tides on Dimune created by its moon, Cressia, when it forms at the same distance our Moon formed (about $\frac{1}{10}$ the distance our Moon is presently) would be only ten times higher than the Moon's tides on Earth today. This may sound like a lot, but keep in mind that when our Moon was that close the tides it created were 1,000 times those we experience now. With the lower tide it generates, Cressia will not spiral out as far as has our Moon. However, there are other factors that will affect its orbit, so I will leave it now and return to it after we have dealt with them.

The length of Dimune's year, the time it takes to orbit Bantam, will be only about 47 Earth days, roughly half the year on our planet Mercury, which is eighty-eight Earth days long. As seen from Dimune, Bantam will cover 1.2° in the sky, 2.4 times wider in angle than our Sun appears from Earth. Although this may suggest that being that big, Dimune will be too hot, the larger size was taken into account along with the star's low temperature in calculating this to be the distance where the temperature is appropriate for liquid water on Dimune's surface.

Let us return to the issue of molten rock tides created by Bantam on the surface of Dimune when the planet is young. We have already noted that the friction created by those tides would slow the planet's rotation. As long as the tides keep the surface moving, this friction will also generate enough heat to keep the surface and most of the interior molten (with the exception of the inner iron core, which will solidify). If you recall Tyran and Mynoa or the Earth and Moon, you probably already know the fate of Dimune: as this latter planet's rotation rate decreases, it will eventually come into synchronous rotation around Bantam. In other words, it will be rotating at the same rate that it orbits the star.

Until Dimune reaches synchronous rotation, its molten surface will release gases from inside it, as did the Earth when its surface was molten and as it does today through volcanoes and mid-ocean rifts. Water vapor (among other gases) will be released into Dimune's atmosphere. It is likely that the amount of water that eventually gets deposited in the atmosphere will be greater than that in the young Earth's air. This will occur partly because the planet will be molten for a long time while its rotation rate is changing and in part because it is closer to its star, hence more comets pulled inward by Bantam's gravity are likely to hit it, depositing their water in its air. This water will remain in the atmosphere as long as the surface is molten because any rain prior to that time will be vaporized before it reaches the planet's surface.

When Dimune arrives at synchronous rotation, the molten rock tides will stop moving along its surface. There will be one high tide directly between the centers of Bantam and Dimune and another on the opposite side of the planet. There will be permanent low tide at ninety degrees from these two high tides. The low tide actually forms a ring around the planet from which Bantam would appear as a big red ball, half of which would be visible on the horizon. Once this situation is achieved, the friction created by the moving tides stops being generated, allowing the planet to start cooling. The crust will then solidify and water will begin condensing into oceans and smaller bodies on its surface.

There is one parameter that I have intentionally avoided specifying so far, namely how elliptical is Dimune's orbit around Bantam. Let's begin with the planet in a circular orbit and reserve the right to consider a more elliptical orbit if a circular one proves too restrictive for the evolution and persistence of life on Dimune.

CIRCULAR ORBIT FOR DIMUNE (DIMUNE I)

As noted above, until Dimune is in synchronous orbit around Bantam, the planet will have neither a solid surface nor liquid water. When the stresses caused by moving land tides end, the surface cools and solidifies, allowing the large quantity of water that has built up in the atmosphere to begin raining down. Because of the initial release of more water than occurred on Earth, the total amount of water on Dimune I's surface will be more than we have. As a result, the continents on Dimune I will cover a smaller fraction of its surface than do ours. This is the same issue that is arising today on Earth as global warming melts ice on Greenland and Antarctica. As this water enters the oceans, they rise, covering more land.

The stage is now set—for disaster. Let's follow the history of water on Dimune I as the planet travels in a circular orbit. Because the planet is in synchronous rotation, the same half of it is continually being bathed by Bantam-light, and the other side experiences continuous darkness. Never getting any heat from the Sun, any water on the continents there will freeze. The oceans on the night side will also freeze, as they have done to some extent in the Arctic on Earth.

Ice is less dense than water, which is why ice floats. Once the surfaces of all the oceans on the night side freeze, water below that ice will freeze, but slowly, as the floating ice insulates it and heat from mid-ocean rifts provides a modicum of heat to the water. Nevertheless, we are talking about billions of years of cooling, which will lead to much of the ocean water on the night side becoming solid. Furthermore, the limited tidal motion of the oceans created by Cressia will do very little to crack up the ice covering the night side of Dimune.

The side of Dimune I in continuous Bantam-light is continually being warmed by that star. As on Earth's equatorial regions, the part of Dimune directly under Bantam is heated most, whereas the low-tide ring of land around the planet where Bantam is seen on the horizon receives the least heat from it. Continuous heating will cause a lot of water to evaporate, especially from the region directly under Bantam. To understand the consequences of this transformation, we need to explore the winds on Dimune I.

Wind patterns are determined by myriad parameters, notable among which are the heating of the land and water, and the rotation rate of the world. The places on the surface hotter than neighboring areas cause convection of the warmed atmospheric gas, which rises upward. This air is replaced by cooler gases adjacent to it. When the warmer gas rises up sufficiently high in the atmosphere (typically between seven and eleven miles above the surface for Earth), it gives off the heat it has and then drifts aside from the rising gas and settles back down toward the planet.

As a planet rotates, the rising and sinking gases are deflected horizontally and vertically, by amounts that depend on their latitudes. This combination of rising and settling, plus deflection due to rotation, is primarily responsible for the different regions of winds. Figure 8.4 shows some of the circulation of the Earth's atmosphere.

Because Dimune I is rotating forty-seven times slower than Earth, the global wind patterns on it would not be nearly as distinct and as powerful as the winds we experience on Earth. The more slowly moving Dimune I just can't drag the air around like Earth can. Nevertheless, some of the moisture raised into the air by the heat from Bantam will travel around the planet as it slowly revolves. This is where the problem begins, because the side of Dimune I in continuous darkness does not receive any heat directly from Bantam. The moisture arriving there in the atmosphere will begin cooling.

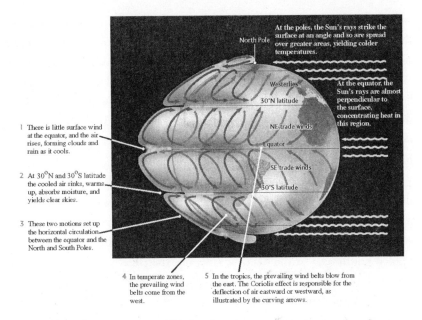

At the poles, the Sun's rays strike the surface at an angle and so are spread over greater areas, yielding colder temperatures.

North Pole

Westerlies

30°N latitude

At the equator, the Sun's rays are almost perpendicular to the surface, concentrating heat in this region.

NE trade winds

Equator

SE trade winds

30°S latitude

1 There is little surface wind at the equator, and the air rises, forming clouds and rain as it cools.

2 At 30°N and 30°S latitude the cooled air sinks, warms up, absorbs moisture, and yields clear skies.

3 These two motions set up the horizontal circulation between the equator and the North and South Poles.

4 In temperate zones, the prevailing wind belts come from the west.

5 In the tropics, the prevailing wind belts blow from the east. The Coriolis effect is responsible for the deflection of air eastward or westward, as illustrated by the curving arrows.

Figure 8.4: Air flow patterns around the Earth. CREDIT: W.H. FREEMAN & CO.

At night on Earth the temperature usually* falls because the heat radiated into space from the air and from the planet's surface is not being replaced by heat from the Sun. The closest analogies on Earth to the permanent night side of Dimune I are the regions above the Arctic Circle during winter in the northern hemisphere and below the Antarctic Circle during winter in the southern hemisphere. The Sun is down for days, weeks, or even months in these two regions during the specified times. The lack of heat and light from the Sun is what makes these regions so cold.

As moist air from the daylight side travels to the nighttime side, it cools and falls as snow. Slow winds and rapid cooling on Dimune suggest that this snowfall will be most pronounced around the planet in the region just into the night side, that is, just beyond where Bantam sets.

Once that snow lands on the night side of Dimune I, it would be extremely

---

* Exceptions sometimes occur because the circulation of the air on Earth is so extensive that warm volumes of air can sweep into places at night and replace the heat the Sun is not providing at that time, thereby keeping the nighttime as warm as daytime.

hard for it to be returned to the daylight side. That return would require that the snow either be vaporized into the air and blown over to the daylight side or liquefied so it flowed downhill to liquid oceans on the daylight side. What little snow and ice on the nighttime side that did get evaporated or liquefied would be more than offset by the constant flow of moisture onto that side from the daytime side. I would expect over millions of years and more that a giant mountain of snow and ice in the shape of a ring around the planet would build up just on the edge of night.

Eventually all the water on the daylight side would be vaporized and then turned to snow and ice on the nighttime side. As this snow piled up, its weight would be so great that it would cause the crust underneath it to buckle and crack, allowing magma out. This would cause some of the ice and snow to melt and return to the daytime side, but the effect would be transient. Eventually the crust would adjust to the added weight of the ice and snow. Then surface magma would solidify and the process of converting all liquid water to ice and snow would go to completion.

The conversion of Dimune I's liquid water to ice is likely to take hundreds of millions of years or more, in part because there is so much water to transform and in part because events such as volcanoes, rifts, and impacts will reliquefy and vaporize some of it from time to time. Indeed, even when the water is essentially all ice, such events will occur and have the same effects. Nevertheless, this process will make Dimune I in circular orbit a particularly inhospitable world. The daylight side will be an arid wasteland, separated from the frigid nighttime side by a mountain ring of ice. There may be a way around this problem of such an uninviting world, namely to put Dimune into an elliptical orbit.

ELLIPTICAL ORBIT FOR DIMUNE (DIMUNE II)

Implicit in Dimune I always having the same side facing Bantam in the scenario above are the facts that the planet rotates at a constant rate and that it has a circular orbit around the star. If the planet were not spinning at a fixed rate, then as it went around, different parts of it would be turned toward Bantam at different times. Altering the rotation rate of such a massive body as Dimune enough to make different parts of it face Bantam on the time scales of a few Earth weeks is not possible without cracking its crust to pieces. In order to maintain some liquid water on Dimune, we need to do something different, namely change the shape of its orbit around Bantam. If the planet, now Dimune II, is in an elliptical orbit around Bantam, then different parts of it will face that star throughout the year. This will occur for two related reasons.

As you can see in Figure 8.5, as the planet changes distance from its star while rotating at a constant rate, different parts of Dimune II do face toward Bantam. Another way to visualize this is to imagine being on Dimune II's equator where Bantam is directly overhead. Find a comfy chair, get a cool refreshing drink, and watch Bantam for a year (forty-seven Earth days). If you were on the previous incarnation, Dimune I, then Bantam would remain forever directly overhead. Now, however, because Dimune II is rotating at a constant rate but changing distance from its star, Bantam will appear to wobble back and forth in the sky throughout each year.

Using the line and arrow on Figure 8.5 for reference, you can see that more than half of Dimune II experiences daylight. We can adjust the amount

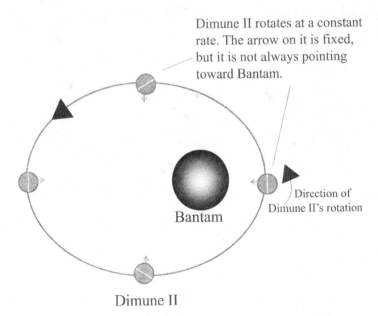

Dimune II rotates at a constant rate. The arrow on it is fixed, but it is not always pointing toward Bantam.

Bantam

Direction of Dimune II's rotation

Dimune II

Figure 8.5: Motion of Dimune II around Bantam. From Dimune II, Bantam moves back and forth in the sky. The white lines on each image of Dimune II separate the daylight from the nighttime sides of the planet, showing that the same side does not always face Bantam.

of the nighttime side that experiences daylight by changing how elliptical the planet's orbit is around Bantam. The more elliptical the orbit, the more of the night side of the planet is exposed to the star. When this occurs, some of the snow will melt and return both to the atmosphere and to the liquid oceans. In this way, there are crescents of land and ocean on Dimune II at the boundaries between the day and night sides where water goes from solid to liquid and back. I call these regions Halbmonds. Let's now consider the water cycle on elliptically orbiting Dimune II.

The vaporizing of ocean water on the daytime side of Dimune II will again lead to snow being deposited on the nighttime side, primarily just beyond the edge of night. The difference with the previous scenario is that after this snow falls on a Halbmond, some of it is liquefied again in the weak Bantam-light. This water runs downstream into the ocean. Snow on the nighttime side that is not exposed to Bantam-light again is mostly locked up and thereby removed from circulation. However, there is one more aspect of this version of Dimune II that will enhance the heat available on the surface for maintaining liquid water. It comes from the land tides on the planet.

In the case of Dimune I, the high land tide closest to Bantam always remains in a straight line between the centers of the two worlds. This is not true for Dimune II. Note in Figure 8.5 that the arrow located at that high tide closest to Bantam does not always point toward the star. Therefore, this bulge of land acts as a handle that Bantam's gravity can pull (like the high tides on the other worlds we have discussed and their attraction to the moons of those worlds). Pulling on that high tide more than on the high tide on the other side of Dimune II creates a stress inside the planet. This stress causes Dimune II to change shape slightly, just as the moon Io does in orbit around Jupiter. Changing shape causes rock inside the world to rub against other rock, which creates friction, which in turn creates heat. Enough heat generated this way, as in Io, can cause rock to melt and come out as lava. For Dimune II, the planet is so large that the extra stresses will add to the internal heat, but won't be enough to make the surface wildly molten (as are parts of Io's surface). Nevertheless, the extra heat generated this way will eventually seep out, as does the heat generated "naturally" inside, so there will be extra heating of Dimune II's surface than otherwise.

Unlike Dimune I, not all the liquid water on this world will be locked up on the nighttime side of the planet. The flow of water from the regions where Bantam is high in the sky to the nighttime side will be countered by water flowing back in the Halbmond regions. These are where life can form and evolve. Let us set the region where liquid water can exist permanently to cover a fifth of the planet's surface, with continental dry land also in that region. The oceans in this region are where life will begin evolving.

Life on Dimune II

The lack of significant tides will delay the enriching of the ocean on Dimune II with minerals necessary to form life. However, the continual melting of snow and the flow of that water in the Halbmond regions will carry down minerals from continents there into the ocean. Therefore, the ocean will eventually have the capacity of forming life.

The energy that enables life to develop in the primordial soup comes primarily from the star as ultraviolet radiation. Because Bantam is so cool, the intensity of ultraviolet radiation from it is roughly 500 times lower than from our Sun. Energy from lightning and volcanic activity will also be available to power molecular activity and, potentially, the formation of life. Nevertheless, the lower level of ultraviolet suggests that the formation of life will be delayed considerably compared to the time at which it began on Earth. Let us say that it begins to occur a billion years later on Dimune II than it did here.

Eventually life transitions from the ocean onto land on Dimune II. When that occurred on Earth, different species migrated over most of the landmasses, sometimes aided by, and sometimes blocked by, the motion of the continents due to tectonic plate motion. Tectonic plate motion will occur on Dimune II. Its effects complicate the activities described here, but do not appear to present any insurmountable obstacles to the existence of life on that planet.

Life on the land of Dimune II will not spread around the globe, as it did on Earth. The frozen, nighttime side of the planet will be far too cold for any Earth-type animal or plant life to survive. Temperatures there will be below −130°F. This effectively precludes nearly half the planet for life except possibly where there is permanent heating on that "dark side" such as on the boundaries of tectonic plates or in continually active volcanic regions such as we see in Hawaii.

The region of the daylight side where the water has vaporized and does not return will be inhospitable arid desert. The temperature in this region will be uniformly above 150°F.*

Life on Dimune II will primarily be restricted to the region from the freezing shores of a Halbmond to the sweltering shores on the other side of its oceans. It will also fill the continents that lie in this region. There would be several profound differences between the biology of the life on Dimune II and the life on Earth. One of these is caused by the fact that there is a much lower level of ultraviolet radiation on Dimune II than on Earth. Ultraviolet radiation from our Sun is responsible for creating the "ozone layer" in our atmosphere. The much lower intensity of ultraviolet from Bantam on Dimune II, coupled with the fact that the ultraviolet that comes from the star is typically less energetic than the ultraviolet that comes from the Sun, indicates that there will be very little ozone generated in the atmosphere of Dimune II. The silver lining for life on Dimune II is that they should be able to live with the lowered intensity of ultraviolet that does come through the atmosphere.

If there is a correlation between the range of wavelengths that eyes can detect and the intensity of the wavelengths of the Sun, then we might expect that eyes on Dimune II will have a different range of sensitivities than do eyes on Earth.† Let's consider why their eyes will probably be sensitive to the same wavelengths we see. Our eyes are most sensitive to yellow light, which is close to the peak of the Sun's colors as seen from the surface of the Earth. Based on blackbody intensity curves, the intensities from Bantam of violet, blue, and green are over ninety times lower on Dimune II than on Earth. Red, orange, and yellow there are some forty times dimmer. Infrared is only about twenty times dimmer on Dimune II than on Earth. It might therefore seem plausible that eyes on Dimune II will peak in what we call the infrared, where the intensity of Bantam's radiation peaks. But this is highly unlikely to occur.

---

* The temperatures cited here are in the realms of the coldest and hottest temperatures on Earth.
† I say "if" because there is active discussion on whether our eyes are synched to the peak sensitivity of the Sun or whether they have the range of color sensitivity they have for other reasons.

The visible part of the electromagnetic spectrum, which we see as the colors of the rainbow, is bounded by the higher-energy ultraviolet radiation on one side and the lower-energy infrared on the other. The energies associated with visible light photons are benign to life, meaning that they don't damage cells. Ultraviolet is so dangerous that even with the best natural protection, many species, such as humans, endanger their vision and skin, among other things, when exposed to too much of it. For example, excessive UV exposure causes cataracts and lesions in the eye that lead to painful photokeratitis. Because Bantam's UV intensity is so low compared to the Sun, there is no compelling reason for nature to evolve sensitivity to it on Dimune II, even if that lower intensity would allow eyes to see it without harm.

Infrared, where most of Bantam's radiation is emitted, presents other problems for detection by "eyes" on Dimune II. We have rudimentary infrared (heat) detectors in our body. If you close your eyes and sweep your hands in front of a fire, you will get a pretty good idea where the fire is located. Creating infrared-sensitive eyes is very difficult for two reasons. First, the wavelengths of infrared photons are longer than those of visible light. Therefore, to have the sensitivity of our eyes to visible light, infrared-sensitive eyes would have to be roughly four times larger than our eyes.

A more pressing problem to evolving these eyes is that the heat generated inside any eye and in its surroundings will emit infrared photons into the eye that will compete with the infrared photons coming into it from the objects being viewed. Consider the light from Bantam striking a tree. Some of that light, mostly infrared photons, scatters off the tree and into the viewer's eye. Those infrared photons entering the eye from the tree have the same wavelengths as photons generated in the eye because it is in a container, such as a head, that keeps it warm. "Warm" means infrared photons. So, unless infrared-sensitive eyes have built-in refrigeration, it will be very hard for the brain to distinguish infrared photons sent your way by objects you view from infrared photons sent into the eye from its immediate surroundings.

It is likely, then, that even on Dimune II, the eyes that evolve there will be sensitive to visible light wavelengths. The "irony" is that even visible light-sensitive eyes on that world will have to be larger than ours in order to collect enough light from its dimmer star, Bantam, to see things as brightly as we see them in the light of our Sun.

The dimness of light from Bantam will also have a noticeable impact on plant life, which gets its energy through photosynthesis. This process takes in light from the Sun or Bantam and uses that energy to manufacture the organic molecules the plant needs to survive. Because Bantam is so dim compared to our star, the leaves on Dimune II are likely to be larger in order to provide the same level of plant activity as occurs on Earth.

9   What If the Earth Had Two Suns?

The train car swayed as it picked up speed, heading north. Ahead, the powerful steam engine marked its passage with a cloud of white smoke and four mournful blasts of its whistle: two long ones, a short one, followed by a final long one. The autos waited patiently as the train rumbled by. The conference on how the Sun shines had ended just hours before, and the train carried twenty of the world's great astrophysicists on the first leg of their journeys home. As always, the conference continued on the train.

"I really don't think it is burning gases," Hans Gethe said, gently, to the young man sitting across from him. "Its spectrum shows virtually no oxygen, which is necessary for combustion."

"But what if the oxygen is deep inside and comes up, perhaps by convection. Before it gets to the surface, it combines with the other gases and ignites," his companion countered.

"I have done a calculation as to how long it would shine under those circumstances. About six thousand years." He smiled, knowing the irony that this number always generated. "Maybe our planet, Zweisonne, and its star were created six thousand years ago after all." He waited, but before the young man could say anything, a middle-aged man approached. Hans recognized Ed Saltwater at once.

"I was very intrigued by your theory that the Sun is actually two stars," Hans said.

Ed nodded. "It would explain why it looks like a lopsided hourglass and why it is brightest at the narrowest point..."

"Brightest because the gases there are compressed most and therefore hottest," Hans interjected.

"Right. Provided that the gases are flowing between the two parts of the Sun. I am working on a hydrodynamic calculation to see which way they would be flowing, if I am correct."

"You know, if you are correct we should have two names for it, um, them, instead of one," Hans observed, with a smile.

"What makes it shine?" the young man asked.

"That, my friend, is the sixty-four-dollar question," Hans replied. "I thought I had an answer before my friend, Ed, here, changed everything with his idea that we have two suns."

Ed Saltwater shrugged. "Beats me right now, too. It's got to be fusion, of course. Like we did in Fat Man and Little Boy."

"I had been working out a scenario where the fusion is taking place in the larger lobe of the Sun. Both lobes, I believe, are rapidly rotating. This rotation generated a magnetic field that bottles up much of the gas in the outer layer of each part of the Sun and causes the gases to transfer from the fusing part to the nonfusing part. The observations of the complex magnetic fields in and around the Sun are consistent with that, but I could not get the total energy generated by fusion in the one lobe to equal the total energy output of the Sun."

"Did you include the energy emitted by the bottled gases?" Ed asked. Hans nodded.

"What if there is fusion in both lobes? If Professor Saltwater is correct, then both stars could have enough mass to allow fusion," the young man said.

The two great scientists stared at him. "What did you say your name was?" Hans asked.

The Earth orbits one star, the Sun, which is isolated by trillions of miles from its nearest stellar neighbors. That need not have been so. In this chapter we explore the possibility and consequences of the Earth forming in a system with a pair of stars bound in orbit around each other. Such binary star systems (introduced in Chapter 8) are quite common. Consider, for example, the brightest star in the night sky, Sirius. It is part of a binary system, with a tiny companion called a white dwarf. This latter body was once a star like the Sun that long ago shed its outer layers as gas and dust into the interstellar medium, leaving only its core of carbon and oxygen orbiting the bright star we see.

Another system provides further insights into binaries. It you look at the middle star of the handle of the Big Dipper, Mizar (Figure 9.1), you will see that there are actually two stars near each other. The dimmer star, Alcor, is close enough to Mizar to create the impression that they are a bound pair of stars, a binary system. This is apparently an optical illusion. With a separation of roughly three light-years, Alcor is about as far from Mizar as the Sun is from its nearest stellar neighbor, Proxima Centauri. Mizar and Alcor may well be unbound; that is, they don't orbit each other. Such a system is called an optical double.

However, if you zoom in on Mizar with a telescope of even moderate

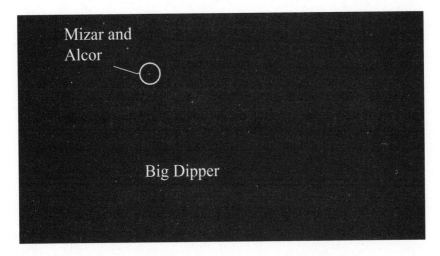

Figure 9.1: Big Dipper showing the optical double of Mizar (brighter) and Alcor.

magnification, you would discover it is, indeed, orbiting with a dimmer companion. This observation was first made in the seventeenth century. The star we see with our naked eyes is called Mizar A, and the dimmer one is called Mizar B. They have quite elliptical orbits around their common center of mass and take about 5,000 years to make one orbit.

In 1889, Edward Charles Pickering examined the spectrum of light emitted by Mizar A and discovered that all the colors were systematically alternating between being Doppler-shifted toward short wavelengths (blue-shifted, meaning that the star was moving toward us) and being Doppler-shifted toward longer wavelengths (red-shifted, meaning that the star was moving away from us). See Chapter 5 for an introduction to Doppler shift. The only way that this alternating Doppler shift could occur is if Mizar A had a binary companion closer to it than Mizar B. As Mizar A and its unseen companion orbit each other, when Mizar A is moving toward us, its colors are blue-shifted and conversely, when it is moving away, they are red-shifted (Figure 9.2).

If that weren't enough, Mizar B was later discovered by the same method to be part of its own binary system. Adding up all the players in this drama, Mizar A is really two stars, Mizar B is likewise, meaning that four stars are orbiting together, and Alcor turns out to be yet another pair. So when you

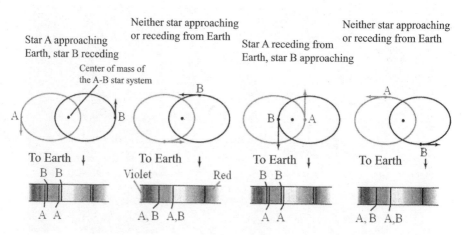

Figure 9.2: Doppler shift: when a star is moving toward Earth, its spectrum is blue-shifted. When a star is moving away from us, its spectrum is red-shifted. When moving across the sky, there is no Doppler shift in its spectrum. CREDIT: W.H. FREEMAN & CO.

look at the middle star in the handle of the Big Dipper and the dimmer star next to it, you are actually viewing six stars.

Roughly half of the objects you see at night as single stars are binary systems or systems with even more than two stars orbiting each other. It turns out that there are many stars near us that are too dim to be seen without the aid of a telescope. Taking these into account, astronomers estimate that about one-third of all the stars in the Milky Way are in binary star systems. In other words, there are "billions and billions" of binary star systems, which raises the question of whether any of them could have life-supporting terrestrial planets and, if so, where might those planets be located.

TYPES OF BINARY SYSTEMS

The stars in a binary system could actually have masses ranging from a quarter solar mass each, to about fifteen solar masses each, while allowing an Earthlike planet to form in their orbital realm. Stars with the upper mass here shine on the main sequence (fusing hydrogen into helium like the Sun) for only around fifteen million years. This is about the lower estimate for how long it would take an Earthlike planet to form. Clearly, a fifteen-solar-mass star would evolve from the Main Sequence (and physically expand and overheat the planet) long before sentient life evolved there.

As a practical upper limit for stars shining long enough to allow life to form on planets orbiting them, the stars would both have to be less than about 1½ solar masses. The two stars Leah and Zeph in the binary system we explore in this chapter have one solar mass and a quarter solar mass, respectively. Based on seeing the Sun in our sky and the discussion of a quarter-solar-mass star in Chapter 8, we know that people on a planet orbiting this system will see one yellow star and one red star. The masses of Leah and Zeph are based on a plausible scenario for an Earthlike planet, which I call Zweisonne, existing in their system.

From the perspective of a terrestrial planet, there are three realms of binary star systems, depending on how far apart the two stars are. We focus on the situation in which the two stars are closer together than the planet's distance from them, but it is worth briefly exploring the other two possibilities.

WIDELY SEPARATED STARS

One scenario is that the two stars are sufficiently far apart that the planet orbits one of them, with the other star being, for example, 100 times farther away. Separated by 100 AU, for example, these two stars orbit each other once every 112 years.

Orbiting Leah (One-Solar-Mass Star)

In order to be in the Habitable Zone of its star, Zweisonne I orbits one-solar-mass Leah, roughly where we are in orbit around the Sun. Located 100 times farther away, Zeph would be some 140,000 times dimmer than Leah. Indeed, Zeph would be barely noticeable as a red star in the night sky. However, it would have a profound effect on the early evolution of Zweisonne I.

Recall from Chapter 2 that there are billions of comet nuclei in orbit around the Sun out beyond Neptune, extending about a trillion miles into space. This realm is, in fact, where Zeph orbits. The gravitational disturbance created by Zeph would cause many of the comet bodies out there to be vaporized as they fell toward that star. Many others would be flung around it into elongated orbits that would carry them toward Leah and Zweisonne I. As a result of the disturbances of the Kuiper belt and Oort comet cloud by Zeph, young Zweisonne I is going to face many more impacts from comets than did the young Earth. These events will, of course, change Zweisonne I compared to the Earth by adding more water to it than we have here, as well as to potentially cause more mass extinctions as these impacts are likely to extend for billions of years into the life of the planet.

If sentient life evolves on Zweisonne I before the comet-clearing process driven by Zeph is over, the tails of the comets near that star could be so bright that Zeph would appear from the planet to have hair. Similarly, comets orbiting Leah would be much more common in Zweisonne I's night sky than they are here. Although it is true that the Earth will be struck by dangerous debris over the next thirty million years, it is statistically extremely unlikely that a piece large enough to cause a mass extinction will strike our planet in the next million years. The odds are much much greater that such impacts would occur more frequently on Zweisonne I. Therefore, developing a warning and deflection system (see countless sci-fi movies over the past decades) would be imperative on that planet, but is less crucial here.

Orbiting Zeph (Quarter-Solar-Mass Star)

If the life-sustaining planet, now Zweisonne II, formed in orbit around Zeph, there would be some noticeable differences from life on Zweisonne I. The orbit of Zweisonne II would be as described in Chapter 8. The gravitational tug from distant Leah would be $1/10,000$ as strong as the Sun's pull or, equivalently, twice Jupiter's attraction to us. This small force would help nudge Zweisonne II into a slightly more elliptical orbit, as described in the previous chapter.

Leah would appear about forty-four times brighter than the full Moon in the sky of Zweisonne II. Leah would also raise the temperature on Zweisonne II an average of about 6°F. When the two stars are on different sides of the planet, the frigid nighttime side of Zweisonne II would be illuminated (and heated) by Leah. The extra 6°F is not a lot of extra energy, but it would cause enough extra evaporating and melting of ice for the evolution of the surface and of life on that planet to be more dynamic than what happens to Dimune in Chapter 8.

It is worth noting that Zweisonne II will also face a barrage of comets scattered from their orbits by Leah just as Zweisonne I faced as a result of Zeph's motion, discussed above. Because Zweisonne II is closer to its star, Zeph, than Zweisonne I is to Leah, Zweisonne II is likely to be hit more often by comets that pass near to Zeph. Most of those impacts will be on the night side of the planet. They will vaporize ice and break up the icy surface there, providing more liquid water for life to evolve on the daylight side, as discussed in the last chapter.

Orbiting Both Stars

There is an intriguing third possibility for inhabitable planets formed in a binary star system with widely spaced stars. Planets could form around both stars. Sentient creatures on both worlds becoming aware of each other would have interesting sociological consequences. When telescopes are available, they would be able to see the effects each has on their planet. When radio technology is developed, they would be able to communicate. And when space travel is perfected, they could visit each other. There are innumerable fascinating sociological, political, religious, technological, medical, and, alas, military possibilities in such interactions.

MODERATELY SEPARATED STARS

When Zeph and Leah are separated by distances in the range ¼ AU and 50 AU, they are unlikely to form terrestrial planets. Recall that planets form from disks of gas and dust orbiting stars as they form. The gravitational tugs of a second star in close vicinity to it are likely to disrupt such a disk orbiting either star before it has time to create protoplanets and planets.

In the unlikely event that a planet does form when Zeph and Leah are, say, 25 AU apart, its fate is not promising. The combined gravitational attractions of the two planets will cause it to move in a very complex orbit that will eventually give it enough energy to either fall into one of the stars or to be ejected from the system.

There is a variety of parameters we should acknowledge that affect the presence and properties of planets orbiting widely separated and moderately separated stars. In addition to the distance between the stars, their masses, and their surface temperatures, these include at least: the angle between the plane in which the planets orbit and the plane in which the two stars orbit each other; the masses of the planets; the number of planets; the ellipticity of the orbits of the stars (i.e., how close they get and how far apart they get); and the numbers of comets in orbit around the stars. Although some combinations may allow a life-sustaining planet to exist in orbit around moderately separated stars for the billions of years necessary for life to evolve on it, such orbits are likely to be too rare to justify elaborating on them here.

CLOSELY ORBITING STARS

The scenario that unfolds when Zeph and Leah are so close together that Zweisonne III orbits both of them is particularly interesting. In this case, the two stars are seen to be near each other at all times as viewed from the planet. I assume that the two stars are in a typical very-close binary orbit that brings them around each other once every Earth day. That would put the two stars in the binary 0.02 AU apart.

Zweisonne III will be identical in physical properties to the Earth, including having a Moon-sized moon. The length of the day on the planet will be twenty-five hours long when sentient life evolves there. I set the disk of gas and dust from which Zweisonne III and any other planets develop to be in the same plane as the two stars orbit each other. Because Zweisonne III is in this plane, the two stars will eclipse each other as seen from that world (Figure 9.3). The total amount of light (and heat) cast upon the planet by the two stars therefore varies cyclically over a one-day period. We can get a feeling for the effects of the eclipses by starting when Leah and Zeph are side by side in Zweisonne III's sky.

With both stars shining, the sky brightness would be only slightly brighter than the sky we see today. This is true because Zeph, which has the same properties as Bantam in Chapter 8, provides only one percent more heat and light than we get today from our Sun. Let time advance until little Zeph is in front of Leah as seen from Zweisonne III. The smaller star, which has roughly a third the diameter of the larger one, would block thirteen percent of Leah's radiation. The brightness of the pair as seen from the planet would drop by more than ten percent for nearly half a day. Then the brightness would return to maximum when the two stars are again side by side.

When Zeph goes behind Leah, all of the smaller star's light and heat is blocked from Zweisonne III. Because Zeph is intrinsically cooler than Leah, this eclipse will change the light and heat arriving at Zweisonne III by much less than during the previous eclipse. The length of the eclipse of Zeph will be the same as when Zeph blocks part of Leah. When Zeph

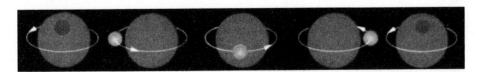

Figure 9.3: Eclipsing binary star system. The larger body is Leah, the smaller one is Zeph.
CREDIT: W.H. FREEMAN & CO.

comes out from behind Leah, the sky of Zweisonne III would return to maximum brightness and heat.

The distance of Zweisonne III from the pair of stars is determined, as usual, by the requirement that the planet be in the Habitable Zone of the system. It may seem plausible that because the side-by-side output of Leah and Zeph is slightly greater than that of the Sun, we should have the planet at or slightly farther than the Earth from the Sun, but actually we want to put Zweisonne III closer to its stars. The reason becomes clear as we proceed.

I place the planet at 0.9 AU. This is slightly inside the most conservative calculation for the inner edge of the Habitable Zone; however, it is likely that a world here will be inhabitable when more details of planetary activity, such as cloud cover, are taken into account.

The added effects of changing brightnesses and temperatures due to the eclipses, although different from what we experience, present no obvious insurmountable problems for the formation and evolution of life. Because the total mass of the stars holding Zweisonne III in orbit is greater than the mass of our Sun, the length of a year on Zweisonne III is shorter than our year, namely 279 Earth days long.

In our solar system, planets and space debris such as asteroids and comets all have stable elliptical orbits around the Sun because our star is a single attracting body (ignoring the effects of other large, but distant bodies in the system, such as Jupiter and Saturn). With Leah and Zeph both pulling on planets like Zweisonne III in slightly different and varying directions, the orbits in that system will not be precisely elliptical, which raises the question of whether Zweisonne III is in a stable orbit.

To understand a little more about the stability, note that objects in Leah and Zeph's star system orbit around the center of mass of the two stars. When the stars are lined up one behind the other as seen from the body they are pulling, that point is directly toward their centers. However, when they are side by side, the center of mass is between the two stars, closer to Leah. When matter is close enough to the pair, the changing location of the two stars creates periodic tugs on the matter. For example, when the stars were forming, their periodic tugs would have pulled rubble that formed certain objects, such as Mercury, out of orbit, preventing that planet from forming. Similarly, when comets fly close to the stars, they can either give the comets a boost in energy, flinging them out of the system forever, or change their orbit and cause them to fall into one of the stars and be vaporized.

Zweisonne III is sufficiently far from the two stars that it is in a stable orbit. The changing gravitational tugs of the two stars are not great enough to cause the planet's orbit to degrade so as to prevent life from developing there.

In the spirit of keeping the Zweisonne III system as similar to our solar system as possible, that planet will acquire a Moon like ours. At the equivalent of today in Zweisonne III's system, the tidal forces acting on the planet from its moon will be the same as from our Moon on the Earth. But the difference in distance to the stars plus the difference in mass between Leah and Zeph on the one hand and the Sun on the other cause the tides created by Leah and Zeph to be slightly higher than those created by the Sun on the Earth. This increase will create more friction between the oceans and the land on the planet, slowing it down more than the Earth has slowed. That is why the day on Zweisonne III today would be about twenty-five hours long, as mentioned earlier.

The Stars and Their Magnetic Fields

The most profound differences between life on Zweisonne III and life on Earth would result from the proximity of Leah and Zeph to each other. Consider some basic traits of these stars. Stars, as we discussed in Chapter 5, are gaseous from their centers to the outer reaches of their atmospheres. When two stars are as close to each other as these two are, the gravitational forces from each greatly affect the gases of the other. In the first place, the tidal force from each star on the other creates a pair of high tides, just as the Moon and Sun create high tides on Earth. Therefore, Leah and Zeph are going to be oval-shaped.

If the Sun is typical, most stars are rotating when they form. Our star spins around about once a month. Unlike the Earth and other terrestrial planets, the different parts of the Sun at different latitudes rotate at different rates. The equatorial regions rotate fastest, going around once every 25.4 days, with the gases at 30° north and south latitude going around once every 26.3 days, and at 60° they go around once every 28.4 days. This variation is called differential rotation.

At the same time, the gases just below the Sun's visible surface, its photosphere, are convecting (see Chapters 4 and 5). The surface is simmering, like a pot of soup. Recall that this convection occurs because the energy generated in the Sun's core is able to heat the gases just below the photosphere, creating hot regions of gas that rise, like bubbles, relative to the cooler neighboring gases.

The gases just described are so hot that the nuclei of atoms (containing protons and neutrons) separate from the electrons orbiting around them, creating a gas called a plasma. As plasma moves, it creates magnetic fields. The Sun's rotation guides these fields into the same shape as those of a bar magnet (Figure 5.4). The complicating factor is that the differential rotation of the Sun causes the magnetic fields to stretch (Figure 9.4).

Just as a rubber band gains energy when it is stretched, magnetic fields gain energy as they are elongated by the differential rotation and by the up and down motion of the convecting gases. Unlike rubber bands, however, magnetic fields don't tear when they get stretched too tight. They always maintain complete, albeit jumbled-up, loops. A star's magnetic field loses the energy it gains when stretched by emerging from below the photosphere and untangling itself. In our solar system, such events have numerous effects on both the Sun and the planets.

Wherever the fields leave and enter the Sun, they repel the gases, creating regions on the photosphere that have less gas in them than the surrounding area. Less gas implies less energy in those regions where the photosphere is

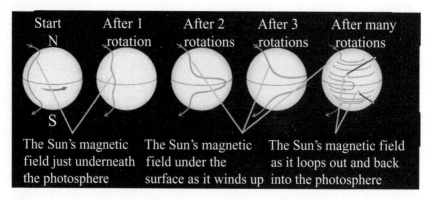

Figure 9.4: The Sun's differential rotation stretches its magnetic fields. CREDIT: W.H. FREEMAN & CO.

pierced. Less energy implies that those regions are cooler than the surrounding gas. Because the surface of the Sun is a blackbody (as introduced in Chapter 8), the cooler regions are darker than the normal regions of the surface; we call them sunspots.

Accompanying sunspots and the magnetic fields that create them are a number of energetic events that sometimes directly affect the Earth. The two most powerful ones are solar flares and coronal mass ejections, mentioned briefly in Chapter 2. Solar flares, usually connected with sunspots, occur as magnetic fields release energy. This release causes the ejection of high-energy particles through the sunspot accompanied by all kinds of electromagnetic radiation, from radio waves to gamma rays. Coronal mass ejections are huge bubbles of gas that are released through the Sun's outer atmosphere, the corona, by magnetic fields. They carry into interplanetary space billions of tons of gas traveling at very high speeds.

Each flare or coronal mass ejection heads outward from the Sun or other stars in a limited range of directions, like shotgun blasts. If you are not in the region of space through which it passes, you can see it, but it will have little impact on you. If, however, you are in its path, its effects can be catastrophic. Astronauts in space struck by the particles and radiation from either event are likely to die. The particles and radiation from them also damage satellites, overload power lines on Earth, and cause magnificent auroras. We have so much money and technology invested in space that we now have space weather satellites that monitor the Sun to give us warning of when a flare or coronal mass ejection is likely to come our way. This knowledge can give us time to put satellites in relatively safe modes, as well as prepare the electrical grids on Earth for the flow of energy that is going to hit them from space.

The final aspects of the magnetic fields of stars that we need to know about, in order to understand major differences between events on Earth and Zweisonne III, relate to the winding up of the magnetic fields. Because the equatorial region of a star rotates fastest, the magnetic fields inside it tend to concentrate toward the equator as time goes on (Figure 9.4). As a result, the locations of sunspots start forming at latitudes around 30° north and south, and as time goes on, new ones form closer and closer to the equator. In the case of our Sun, after about eleven years of sunspots forming, the magnetic fields release sufficient energy (get sufficiently untangled) so that they sink back into the Sun, clearing the surface of sunspots. Typically within two years another sunspot cycle begins with new sunspots forming in those mid-latitudes.

Now let's consider how magnetic fields and the resulting activity on the stars will be different in the Leah–Zeph star system with the two stars both rotating and being so close to each other. The first difference between the Sun and this binary system originates in the oval shapes of the stars created by the tides on them.

As the surface gases of each star rotate, they change distance from the center of their star. This will cause the convection on the surface to change dramatically from what occurs on the Sun. The simmering gases will not be uniform all the way around each star. Like the pistons in a car motor, the rising and sinking of the gases will compress the gases below them by different amounts. When the gases are at high tide (between the two stars and on the opposite side from the other star), they will be spread out and cooler, whereas at low tides, the gases will be compressed and hotter. Therefore, each star will not appear uniformly bright. They will have hot spots at the low tides that are brighter than the rest of the star, along with cooler spots at the two high tides they each have.

This variation in temperature also implies a variation in color. You can understand this by watching an electric stove heat up, as discussed in Chapter 8. Based on Wien's displacement law, first worked out in detail by Wilhelm Wien, and the Sun's color, Sunlike Leah will appear orange at its high tide locations and white at its low tides. The cooler Zeph will appear deep red at high tide and slightly orange at low tide. Between these two extremes on each star, the colors will change smoothly.

Although the coupling of gravity between Leah and Zeph has another major role to play, let's first explore the interaction generated by their magnetic fields. The two stars were formed from the same fragment of a cloud of gas and dust. As this debris came together under its mutual gravitational attraction, it formed a disk rotating rapidly enough that the two stars were formed together, along with the disk from which the planets, including Zweisonne III, formed. In particular, the two stars formed with rotation axes in the same direction, perpendicular to the plane in which they orbit, and both spinning in the same direction. The consequence of this plausible formation scenario is that both have north poles on the same side and like

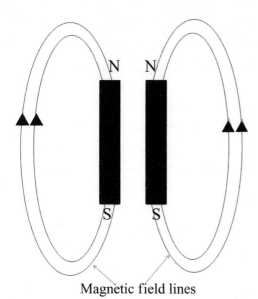

Figure 9.5: *Like two parallel bar magnets (shown here), the magnetic fields of Leah and Zeph repel each other, leading to complicated motion of gas in the atmospheres of the two stars.*

Magnetic field lines

two bar magnets aligned parallel, they have their magnetic north poles adjacent to each other. The interaction between the two fields, although quite complicated, is basically as shown in Figure 9.5.

There is a significant complication to this discussion of stellar magnetic fields. For reasons that are still being worked out, the direction in which a star rotates does not, by itself, determine the direction in which its barlike magnetic field points: if the Sun is typical, the magnetic axis of a star periodically flips. In other words, the north magnetic pole becomes the south magnetic pole and vice versa. In the Sun's case, that occurs every eleven years. This is likely to happen in Leah and Zeph, but they are also very likely to flip magnetic axes at different rates. As a result, they will sometimes have opposite magnetic poles next to each other.

The distortions of the magnetic fields created by the proximity of Leah and Zeph will greatly affect the dynamics of flares and coronal mass ejections. Considering that the understanding of our Sun's magnetic field and its ability to generate events such as flares and coronal mass ejections is still a work in progress, I posit that the coupling of Leah's and Zeph's fields, along with the tidal change in the distance between the surfaces and centers of each star, will lead to even more ejection activity than we experience in our solar system.

During the times when the magnetic fields of the two stars are inverted relative to each other, gases will flow more freely from one star to the next, channeled by the magnetic fields. This will greatly complicate the ejection activity and, in all likelihood, make the Leah–Zeph system much more active and dangerous than when the fields are parallel.

The consequences to Zweisonne III and its astronauts of the more frequent and more energetic stellar storms than we experience from the Sun will be significant. Zweisonne III will experience many more auroras, of course, and space travelers will be in greater danger from the high-energy particles that leave the stars. One intriguing consequence of the increased activity from Leah and Zeph relates to the power lines carrying electricity around Zweisonne III. The Earth and Zweisonne III are surrounded by magnetic fields that create Van Allen belts around both worlds (see Chapter 2). These fields are disturbed when struck by high-energy, electrically charged particles from their respective stars. The changing magnetic fields, in turn, generate electrical currents in the power lines that carry electricity around the planet. These errant currents can destroy electrical equipment, thereby causing power blackouts, as has happened on Earth on a number of occasions. One notable such event occurred in 1989, when currents induced by a solar storm burned out a transformer in the HydroQuebec power system. In less than two minutes this event led to a blackout that lasted over nine hours and affected over six million customers.

Because Zweisonne III will be subject to more flares and coronal mass ejections, its power grid will be more at risk than ours. Engineers and scientists there will have to design a more resilient electrical network than is necessary on Earth.

The more frequent bombardment of high-energy particles from Leah and Zeph will also have a significant effect on Zweisonne III's atmosphere. Recall that the normal flow of particles from the Sun and by extension from Leah and Zeph are trapped by Van Allen belts surrounding planets like Earth and Zweisonne III. However, particles released in flares and coronal mass ejections are moving so fast that they traverse the Van Allen belts and strike the upper atmosphere of the planet. The impact of those energetic particles on air molecules heats the atmosphere, causing some of it to leak into space, never to return. Because they occur more frequently from Leah and Zeph, flares and coronal mass ejections will cause even more of Zweisonne

III's atmosphere to leak out into space. Over billions of years, this could measurably decrease the density of the atmosphere there compared to ours.

Roche Lobes and Stellar Evolution

Many of the differences we have seen between the environments of Zweisonne III and Earth are matters of degree. However, one stands out as being profound. That is the evolution of Leah and Zeph compared to the evolution of the Sun. The difference starts in their atmospheres.

Stars evolve primarily because they convert lighter elements into heavier ones via nuclear fusion. In this process, the nuclei of atoms, composed of protons and neutrons, are able to fuse because they have enough energy to smash together. When their energies are too low, the repulsion that the protons in each atomic nucleus exert on the other prevents fusion. As introduced in Chapter 5, the energy that atoms have is measured by their temperature. The hotter they are, the faster they move. For our purposes, we need only focus on the fusion of hydrogen into helium in stars like the Sun, Leah, and Zeph.

The minimum temperature necessary for hydrogen fusion to occur is roughly 10 million K. The only region in these stars where that temperature is exceeded is in their very centers, their cores. To generate such high temperatures (room temperature is about 300 K), gases in a star must be compressed by the gravity of the entire star's mass pushing down on its core (whenever gases are compressed they heat up). Both Leah and Zeph have enough mass to reach this 10 million K threshold, but because Leah is more massive (same mass as our Sun), more of its core is hot enough to fuse than in Zeph (a quarter the Sun's mass). Therefore, Leah is brighter and hotter than Zeph, as discussed earlier.

The photospheres of the two stars in our binary system are only two million miles apart. We have already seen that their proximity to each other leads to enormous tides. The question then becomes: can the gravitational attraction of one of them pull some of the gases from the other onto it? That is, can one of them become more massive, at the expense of the other? This question came to the attention of Édouard Roche, who derived the equation showing how the atmospheres of two stars as close as Leah and Zeph behave.

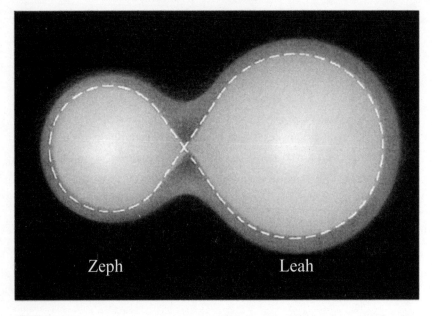

Figure 9.6: *The atmospheres of the close binary stars Zeph and Leah change shape and overflow their Roche lobes (dashed lines).* CREDIT: W.H. FREEMAN & CO.

An isolated star like the Sun has a more or less spherical atmosphere* held in place by the star's gravitational attraction for that gas. When two stars are as close as Leah and Zeph, their combined gravitational fields change the shapes of the atmospheres to overlapping teardrops (Figure 9.6).

As long as they remain inside a boundary called the Roche lobe, the gases in the atmosphere of a star stay bound by that star's gravity. Once outside the Roche lobe, the gas can be attracted onto the other star. This last point is the key to what is going to happen to Leah and Zeph. Some of Leah's atmosphere, which extends farther from it than does Zeph's atmosphere, will leak out of the Roche lobe surrounding Leah at the point where the lobes from the two stars meet. As a result, this star will give mass to Zeph.

As Leah's mass decreases, the force of gravity that compresses this star will also decrease, which in turn will decrease the pressure and temperature in its core, decreasing the amount of fusion that occurs in it. Leah will there-

---

* Effects that make it nonspherical include magnetic fields and their related phenomena, as well as rotation.

fore dim and become redder over time. At the same time, Zeph's mass increases, hence the pressure and temperature in its core both increase. Zeph will fuse more than it did at first, becoming hotter and brighter.

Besides changing brightnesses (luminosities) and colors, the two stars will spiral in toward each other as mass from Leah transfers to Zeph. Therefore, their orbital period will decrease from its initial one orbit per day. As they get closer, the rate of transfer between them will increase until they both have equal masses of $5/8$ solar masses. Although the details have yet to be worked out, it is likely that when they are both of equal mass there will be transfer from one to the other lobe for a while. Depending on such details as the spin and magnetic fields, this transfer will probably reverse, perhaps leading to a cycle of transfer first to one star, and then back to the other.

Both stars change luminosity as they transfer matter from one to the other. Overall, the total output of energy by the two stars decreases as their masses initially become more equal. The change in brightness and, hence, the change in heat from its stars will have a perceptible, but probably not catastrophic impact on Zweisonne III. Two factors go into this statement. Most important is that it will take hundreds of millions of years, possibly longer, before the two stars have equal masses. Their total energy output will then be less than that of the Sun. That is the reason I put Zweisonne III closer than 1 AU from the system in the first place; it will still be in the Habitable Zone of two-equal-mass Leah and Zeph after this transformation occurs.

When life evolves on the planet, they will see in the sky not one star, but two, surrounded by hot gas, with the whole system creating a glowing figure-eight pattern. If the mass transfer oscillates, as mentioned above, then that pattern will be lopsided for much of the time. As noted earlier, because these stars orbit each other in the same plane as Zweisonne III orbits, people there will see frequent eclipses. Noting above that the stars get closer in the process of exchanging mass and that they will be equal-sized, by the time sentient life watches them, the eclipses will cause the brightness of the sky to change by half every few hours. That is, when the stars are side by side, the sky will be twice as bright as when one star eclipses the other. This is, of course, very different from the situation when the two stars had just formed, described earlier in this chapter.

There is another benefit for Zweisonne III to decreasing the mass of Leah.

Stars with lower masses shine longer before they expand and shed their outer layers. Our Sun, for example, will have shone pretty much as it does today for a total of about ten billion years before expanding and making Earth uninhabitable. The Sun is about halfway through that period of stable shining now. With a mass decreased to $5/8$ solar mass, Leah and Zeph will each be able to shine for thirty-two billion years before expanding. Therefore, all other things being equal, Zweisonne III will be inhabitable for a longer time than will the Earth.

What If Another Galaxy Collided with the Milky Way?

EARTH

"It's going to hit the Sun," the computer said.

"'It' what, the whole cluster of stars is going to hit the Sun? Come on. You can do better than that," Sam said, with an edge in his voice.

"I'm sorry, Sam. I should have said that star 21343 in the globular cluster M2 is going to hit the Sun."

"I feel better already," said Sam. "Now tell me about the collision."

"M2 has 154,723 stars plus a black hole of 1,327 solar masses. It has been coming our way for 37 million years, having entered the disk region of our galaxy a little over 1.2 million years ago. Qualitatively, you would say it is coming up behind us."

"Since when have you been qualitative?"

"Since my latest software was installed. Didn't I use the correct descriptor?"

"Yes," Sam said, with a sigh, "but whoever upgraded a scientific computer with qualitative descriptors should be vaporized. Anyway, carry on."

"The black hole in the cluster will pass ninety light-years from Earth, posing no danger. There will be eighteen stars that pass within five light-years. Among these is number 21343."

"What do you know about that star?"

"It is a K0 Main Sequence star, meaning that it has three-quarters the Sun's mass, eighty-five percent of its diameter, and a surface temperature of about 5,100 K, 800 degrees cooler than the Sun."

"Does it have any planets?" Sam inquired.

"No. That is consistent with the fact that its rotation rate is less than once a century, suggesting that it did not have a gas disk around it when it formed."

"No gas disk means no planet formation."

Sylvia, the computer, nodded in agreement.

"When will it hit?"

There was a momentary pause, followed by, "161,828 years."

"Have you told this to anyone else?" Sam inquired.

She shook her head.

"How do you feel about this, Sylvia?"

"Quite honestly, I feel like running around shouting, 'The sky is falling; the sky is falling,' but seeing how you're taking this makes me feel better." She paused for two heartbeats. "Tell me, Sam, why are you so sanguine?"

"Before I tell you, I want you to run another simulation of the next 161,828 years. Your simulations use what, sixty-four significant figures?"

"That's right."

"Okay, I want you to use an initial mass for the black hole of 1,327.00000000000001 solar masses. Keep everything else exactly as before."

"Instead of precisely 1,327?" Sylvia asked.

"That's right. In the meantime, I'm going for a cup of coffee."

An hour and a half later Sylvia paged Sam, who was in the middle of a retro computer game called "Lunar Lander." Finishing a successful landing and earning his prize at the lunar McDonald's, he wandered into Sylvia's lab.

"And?" he asked.

"With the identical run, except for that tiny change you made in the initial conditions, none of the stars in the globular cluster hit the Sun."

She looked confused, but after studying him for a second, added, "You knew that would happen, didn't you?"

"I didn't know, but I confess that I expected this result. I'd like you to make another run, changing the 1 you added in 1,327 to a 2."

When this simulation was completed, Sylvia called Sam on his cell phone. "There wasn't a collision, but there was a near miss between the Sun and another cluster star, which missed it by only 0.04 AU. That is about ten times closer than Mercury is to the Sun. At that distance, the Sun would be shaken up very badly and all the terrestrial planets would be sent into highly elliptical orbits or even freed from the Sun entirely."

Sam nodded, thoughtfully. "As I suspected. Despite your very advanced algorithms for calculating the motion of the stars in that globular cluster, mathematical chaos is still at play. Even the slightest change in the initial conditions generates profound changes by the time the simulation is through. And if the observational astronomers have made even the slightest mistake in the positions or velocities of the stars in that cluster, then we will never get it right.

"So, the best we can do is to have you run 10,000 slightly different scenarios and we will then average the result to get a probability that the Sun will be destroyed or that the Earth's orbit will be changed so much that our planet will be unable to sustain life."

"That is what you will publish, a probability?"

"Yup. A lot of science is probabilities."

Thus far, we have explored the consequences of changes to an Earthlike world, to its immediate astronomical environment, and to its location in space and time. Each chapter culminates in a more or less inhabitable world that is different from our own. In this final chapter I want to step back and explore the consequences of a "What if?" scenario that affects both the Earth and its entire home galaxy, the Milky Way. What if a galaxy collided with the Milky Way? How likely is it that a star from the intruder will hit the Sun or the Earth? This chapter explores an event that has significant impact on our world because it is actually going to happen. We begin by examining how it is that our solar system has remained undisturbed by stellar intruders for over 4½ billion years.

Most of the stars in our neighborhood of the Milky Way move in nearly parallel paths to the Sun around the center of our galaxy. As a result the probability that one of these stars will ever be on a collision course with the Sun or even come close enough to disturb the Earth's orbit is slim to none. There is just too much space between stars. A close encounter between the solar system and another star is much more likely to occur if a large number of stars were initially moving toward us, rather than moving parallel to our motion. The greatest possibility for such an event is in the collision between the Milky Way and another galaxy.

Galaxies like the Milky Way are not isolated travelers in the universe, as we saw in Chapter 5. We orbit with about sixty other galaxies in a gravitationally bound ensemble called the Local Group, which is dominated by two large barred-spiral galaxies, our Milky Way and Andromeda. Surrounding both of these galaxies, like black flies in summer, are dozens of smaller, less massive galaxies. In this chapter we explore the consequences of a collision between the two major players in the Local Group.

Observations reveal that Andromeda is likely to collide with the Milky Way some three billion years from now. The spectra of all the stars we observe in Andromeda are Doppler-shifted toward the blue, meaning that this galaxy is heading our way. At the same time, of course, it is possible that Andromeda is also moving somewhat perpendicular to us, like a car passing from behind, so that it would go by either above or below our galaxy, thus avoiding a collision. It does not appear that we will be so lucky. Calculations based on these observations support the supposition that a collision will occur.

The two galaxies in this scenario are each filled with hundreds of billions of stars and other nasty stuff that can go bump in the night. Furthermore, the stars in Andromeda are most definitely not traveling parallel to us in our orbit around the center of the Milky Way. Although we don't know for sure in what direction such a collision between these galaxies would occur, it is certain that some of Andromeda's stars would be coming toward our solar system. In order to get a handle on what our descendants are going to see and experience, let's explore our galaxy and how the bits and pieces in it and those in Andromeda are going to interact when they collide.

As depicted in Figure 7.1, our Milky Way is a barred spiral galaxy with two major spiral arms, several minor ones, and a bar of stars, gas, and dust running through the middle. The spiral arms orbit in a plane that is filled with interstellar gas and dust. The arms are regions in this interstellar debris where new stars are forming. More about this process was presented in Chapter 5, when we explored the hypothetical galaxy the Musketeer.

From the side, both the Milky Way and Andromeda look like old-fashioned flying saucers, with a thin disk coming out from a central bulge, top and bottom (Figure 10.1). Surrounding the disk of each galaxy is a spherical region called the halo, which contains globular clusters of stars, as well as isolated stars. For example, the Milky Way contains at least 150 globular clusters, each typically consisting of hundreds of thousands of stars bound together by their mutual gravity. Many of these clusters also contain a black hole with a few thousand times the mass of our Sun. As they orbit, globular clusters sometimes cross through the disk of the galaxy, passing the stars and interstellar debris in it. There is a tiny chance that our solar system will

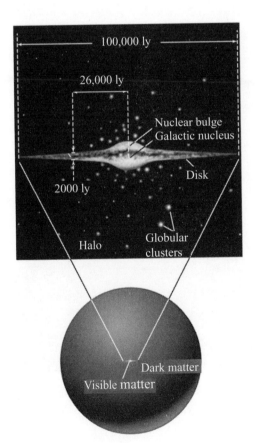

Figure 10.1: Side view of the Milky Way Galaxy. Andromeda has the same profile. CREDIT: W.H. FREEMAN & CO.

someday encounter such a cluster. The effects of such an interaction would be a small-scale version of the collision at the heart of this chapter.

Surrounding the disk of each of these two galaxies and extending far beyond the halo of globular clusters is a vast quantity of matter whose nature we do not yet know. It is presently invisible to our technology; hence it is called dark matter. We know dark matter exists because among other things its gravitational attraction changes the orbits of stars that we can observe, especially in the outer reaches of the spiral arms, compared to how they would track without its presence and influence. There is over five times as much dark matter in our galaxy and, by extension, in the universe, than there is visible matter.

Andromeda is one of four galaxies visible to the naked eye, along with the Triangulum Galaxy and the Large and Small Magellanic Clouds. Located over the Earth's northern hemisphere in the constellation of Andromeda, the

Figure 10.2: Andromeda Galaxy. The oval blob of stars below and to the right on this image is the companion galaxy NGC 205. COPYRIGHT © 2006 BY DR. ROBERT GENDLER.

Andromeda Galaxy looks like a fuzzy blob to the unaided eye. With a telescope it resolves into a beautiful barred spiral (Figure 10.2), like the Milky Way. Until recently, data suggested that Andromeda was the most massive galaxy in the Local Group, but recent observations indicate that the Milky Way has slightly more matter than does our companion galaxy, making our galaxy the most massive one in our neighborhood of the cosmos. The precise masses are still being worked out.

As you can see in Figure 10.2, Andromeda is neither a face-on nor an edge-on galaxy from our perspective. Over the next two and a half billion years, Andromeda will grow larger and brighter in the night sky. Through a powerful telescope right now, it appears about six times wider across than the Moon. By the time it is one-tenth its present distance from us, it will span one-sixth of the night sky and will be clearly visible to the naked eye.

GALAXY COLLISIONS

Given enough billions of years, collisions are actually quite common among galaxies that are orbiting together. Recall from the discussion in Chapter 7 that both the Milky Way and Andromeda have already collided with other galaxies. There are basically two types of collisions between galaxies: ones in which a much larger galaxy collides with a smaller one and ones between galaxies that are nearly equal in mass and size. A collision between unequal systems is called galactic cannibalism because the larger galaxy consumes the smaller one. Both our Milky Way and Andromeda are cannibals. The Milky Way, for example, is presently devouring the Sagittarius Dwarf Elliptical Galaxy. Andromeda apparently broke up at least one galaxy in the past, with fragments still orbiting around it.

The collision that astronomers envision between the essentially equal Milky Way and Andromeda galaxies differs dramatically from the consuming of one galaxy by another. The encounter is going to be more of a merger, but with a twist. The merger has a variety of notable epochs, all of which have potential consequences for the Earth. As we go through the major events in the merger, I will envision snapshots of what will be seen from Earth at those times.

Dark Matter Interactions

Andromeda is heading toward us at a speed of about 270,000 mph. The collision speed will become even higher as the galaxies accelerate toward each other. Recall that dark matter extends far beyond the visible edges of both the Milky Way and Andromeda. As they approach each other for the first time, the first parts of the galaxies that interact will be the dark matter in their halos. Because we don't know of what dark matter is composed, there is little we can say with certainty about what will happen during this time. The best we can do is look at other interacting systems of galaxies to see what happened to their dark matter halos.

Although we can't see the dark matter, we can see some of its gravitational effects, including the way it changes the direction of light passing near it. Called gravitational lensing, this occurs because the dark matter (like all matter) warps the space around it. When passing dark matter, light that would otherwise travel in straight lines actually follows curves like light going through a lens. As a result, astronomers observe distorted images of galaxies beyond regions containing dark matter. The amount of distortion tells us the location and concentration of the intervening dark matter.

Observations reveal that in collisions between galaxies, the dark matter from the separate galaxies appears to merge. After the collision, in most cases the dark matter and the bulk of the visible matter end up occupying the same region of space. However, in at least one case, the resulting distribution of dark matter has become displaced from the visible matter. I choose to have the two realms of matter, visible and dark, remain together for reasons that become apparent shortly. So, to begin the collision, dark matter from both galaxies begins to pile up together between them.

Astronomers looking in the direction of the condensation of dark matter will see the Andromeda galaxy behind it distorted—stretched out—because the light from the stars changes direction as it passes the pile of dark matter. Likewise galaxies behind Andromeda and the dark matter will also appear oddly shaped. Today, when we see such distortions from gravitational lensing of distant galaxies, they often appear crescent-shaped, and sometimes elliptical galaxies appear as hollow rings.

Globular Cluster Interactions

While the dark matter is in the process of merging, some of the globular clusters located in the leading edges of the colliding galaxies will pass near each other or even collide. In either case, their interactions will be messy. Consider a near-miss. As the two clusters pass each other, the tidal gravitational attraction will elongate both clusters (which are spherical today). As they fly by, many of the stars in each cluster will be pulled away from their cluster, scattering into space, leaving two much smaller clusters. Furthermore, their mutual gravitational attractions will cause the two remaining clusters to travel off in very different directions from their original tracks.

When two clusters collide, the effect is even more dramatic. At first, they each elongate due to the tidal force from the other. During the collision, many of the stars in the two globular clusters pass close enough to each other to significantly change direction under the influence of their mutual gravitational attractions. As the two globular clusters emerge from their collision, each cluster will be expanding outward, that is, enlarging. People on Earth will see the remnant of such a collision appear as a glowing bow-tie in the sky.

If collisions between stars are going to occur anywhere in the collision between the Milky Way and Andromeda, they are going to occur in colliding globular clusters. The reason that globular clusters are the best candidates for such events is that these are the densest concentrations of stars in a galaxy.* In their centers, many clusters have a thousand times more stars packed together than we find in our neighborhood. Even compared to the spiral arms of galaxies where stars in the disk are most concentrated, the globular clusters have many times more stars per volume of space.

The evolution of a collision between stars depends on the types of stars involved, how fast they are moving relative to each other, and how close to head-on is the collision. In any event, they are likely to shed their outer layers. If the collision is slow enough and not directly head-on, the inner regions will merge. However, the pressure is so great in the cores of stars that if their outer layers are shed sufficiently rapidly, the cores will spring outward in titanic

---

* Actually, the very center of the galaxy, its nucleus, has a slightly higher concentration and collisions are likely there, too, but this region is very small compared to the sizes of globular clusters and there is only one nucleus per galaxy.

explosions. Such explosions within a few light-years of Earth would be catastrophic for life here.

If pairs of stars pass close enough and slowly enough, they could merge with only minor loss of their outer layers. Even allowing for the loss of mass from the outer regions of each star, their inner volumes are much denser than their outer parts, hence the overall mass of the resulting star will typically be greater than the mass of either of the original pair that collided. As we saw in Chapter 5, stars with different masses evolve differently. Stars that have more than about nine solar masses will explode as powerful supernovae, whereas stars with less mass will end by shedding matter in wimpy planetary nebulae. In order to know whether the Earth might have to face numerous supernovae emanating from a nearby globular cluster (and, it is likely that the solar system would pass near one of these in the collision), it is essential to know what the masses of stars that form from collisions in a globular cluster are likely to be.

We can get a handle on the masses of the collision-formed stars from the masses of stars that will exist in globular clusters three billion years from now. Globular clusters formed early in the life of the universe, meaning that most of them have been around for at least ten billion years. These stars have therefore been evolving for much longer, in some cases nearly three times longer, than the Sun. Assuming, plausibly, that nearly all the stars in a globular cluster formed simultaneously, and knowing how long stars with different masses exist in each stage of stellar evolution, we are able to determine the masses of stars that will still be shining in globular clusters when the galaxies collide. The result is that nearly all the stars in globular clusters three billion years from now will have less mass than the Sun. When any two of these stars collide, the resulting star will have less than two solar masses.* Therefore, the stars that form from mergers in globular clusters will be of low-mass (i.e., less than two solar-mass) stars. Their explosions will be planetary nebulae, so our descendants won't have to worry about potentially life-destroying supernovae, at least from globular clusters.

---

* Yes, it is possible that a star will collide more than once, creating even higher-mass stars. I expect that some will, but these events will be few and far between compared to just a single collision per star.

Fluttering Disks During Interactions

One of the more bizarre things that will occur as the two galaxies approach each other is that their disks of stars, gas, and dust will appear to flutter, like a stingray swimming. Such activity occurs because the side of each galaxy closest to the other galaxy feels a stronger gravitational attraction to it than does the opposite side of each galaxy. The objects on the closer side will rise up toward the other galaxy. At the same time, of course, those stars and that gas and dust are still orbiting around the centers of their galaxies, so after moving up toward the other galaxy, they will move back down as they move away from it. Our descendants will see Andromeda change shape, sometimes dramatically, as the two galaxies approach each other.

Interstellar Gas and Dust Interactions

Before any serious disturbance occurs between them, both galaxies in this collision scenario have huge quantities of interstellar gas and dust (hereafter, gases) orbiting in their disks. Recall that this is the realm in which our solar system moves. In the Milky Way it is estimated that gases account for ten to fifteen percent of the visible matter. As the disks of the two galaxies come into contact, the gases from each will slam into the other at speeds of over 300,000 mph. Such impacts will create huge shock waves through the gases, like sonic booms or the snap of a whip in the air. As a consequence, some of the gas will become extremely hot and will expand away, leaving the gravitational grip of the galaxy in which it had previously been confined. This gas will drift into intergalactic space.

Other gas will be compressed enough by the collision to allow stars to form in it. Considering that the interactions will go on for millions of years, if even a few dozen stars are formed per year as a result of the collision, the entire sky will soon be afire with new stars. If we think in nearly astronomical timescales of millions of years, all this activity will create a burst of stars. The big picture of the colliding gas is that much of it from both galaxies will come to a screeching halt. Likewise, most of the stars formed from this gas will remain where they are formed and the pre-existing stars from both galaxies will cross each others' paths and those of the newly formed stars.

Disk Star Interactions

Like two enormous pinwheels, the disks of the Milky Way and Androm-eda will eventually spiral into each other. As they collide, the gravitational force from each galaxy will pull billions of stars from the other one out of the circular orbits they previously had. These stars will stream into inter-galactic space, creating one or more bright tails (Figure 10.3). Some of these stars will be moving so fast out there that they will leave the vicinity of the galaxies forever, sailing off into intergalactic space.

Most stars, however, will remain bound to the merging pair of galaxies. These stars will fly outward until the gravitational attraction of the other matter in the galaxies stops their outward motion. They will then arc back inward. This is analogous to what happens when you start running and then jump upward on Earth: your path is an arc.

These returning stars will have very different orbits from when they were in the disk of a galaxy. In fact, different returning stars will be in all differ-ent planes of orbit. In other words, the galaxies will eventually lose their

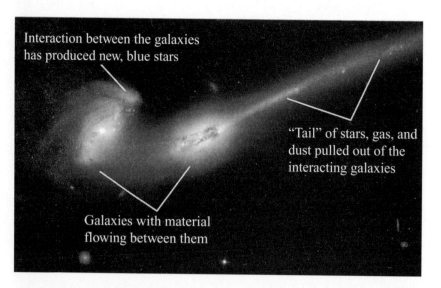

Figure 10.3: Two colliding galaxies with a tail of out-flowing stars, gas, and dust. This system is labeled NGC 4676 (the 4,676th object in the New General Catalog of galaxies). CREDIT: NASA, H. FORD (JHU), G. ILLINGWORTH (UCSC/LO), M. CLAMPIN (STSCI), G. HARTIG (STSCI), THE ACS SCIENCE TEAM, AND ESA.

disks. As noted earlier, collisions between stars in the crossing disks are unlikely because there is so much space between stars in these regions. For example, whereas the Sun has a diameter of 870,000 miles, the closest star to us, Proxima Centauri, is about 25,000,000,000,000 (trillion) miles from us. Therefore the fraction of space taken up by stars is so low that the total number of collisions of disk stars is likely to be at most a handful during the entire galactic collision process.

There will, however, be countless millions of sufficiently close passages of stars that, as with the stars in globular clusters, many disk stars will be pulled into dramatically different orbits than they had originally. This applies not just to the stars that stream out, as described above. Many others will go every which way. As seen from Earth, both the region of light we call the "Milky Way" and the disk of Andromeda will cease to exist. The night sky will be much brighter and the stars will be more randomly distributed.

This encounter between the disks of the galaxies will be the main attraction of the collision. From the perspective of viewers on Earth, it will begin with stars from the outer edges of the spiral arms rushing by our planet. Speaking of bright stars in the sky, it is worth mentioning the patterns of stars we call constellations or, technically, asterisms. As discussed in Chapter 2 the constellations are continually changing shape, even today, because different stars move at different speeds across the sky. During the disk collision phase of this chapter's adventure, the constellations will change more rapidly and more frequently than at any other time in Earth's history. None of the patterns visible today will be visible then.

Eventually it is likely that a star will pass so close to the solar system that the Earth's orbit will be affected. Let's consider a star, Samos, similar to the Sun in mass, passing 100 AU from the Sun, which is roughly twice Jupiter's average distance from it. The outer planets, Neptune, Uranus, and Saturn, will either crash into the star or be thrown out of orbit. Jupiter's orbit will be profoundly altered. The consequences of this event will depend sensitively on the relative speeds of the Sun and Samos, their angle between the Earth's orbit and that encounter, and the location of the Earth in orbit as the passage takes place. The least we can expect is that our planet's orbit will become more elliptical (elongated) and that the plane of the orbit (the ecliptic) will be changed. Samos will be 10,000 times dimmer than the Sun in Earth's sky, which makes the intruder 20 million times brighter than Sirius, brightest star in the sky today.

For the year or so during Samos's closest approach, the Earth's oceans will undergo severe sloshing as the planet adjusts to its new orbit. As a consequence, the shorelines on Earth, and everything near them, will be dramatically affected. Tsunamis of unparalleled height will occur. It is likely that the stresses on the Earth's crust will cause severe earthquakes and stimulate widespread volcanic activity.

The passage of Samos will profoundly disrupt the orbits of the comets in the Kuiper belt and Oort cloud (as discussed in Chapter 2). Many of these comets will be vaporized by the heat from Samos. Many others will be deflected either sunward or completely out of orbit.

Astronomers have begun discovering myriad comets orbiting other stars. Indeed, it is likely that most stars have the equivalent of Kuiper belts and Oort comet clouds. Therefore, many of the comets originally orbiting Samos will be deflected by the gravitational effects of the Sun and planets in our solar system.

For hundreds of thousands of years afterwards, the night sky of Earth will continuously have dozens of comets streaking through it. That period would be a good time for people on Earth to invest heavily in an early warning system and ways of deflecting comets heading earthward.

Although the passage of Samos will be the closest approach of another star to Earth, there will be many other important gravitational interactions between stars from Andromeda and our solar system. A few stars much more massive than the Sun will pass close enough to drastically change the orbit of the Sun from its present nearly circular orbit around the nucleus of our galaxy. Instead, the solar system will be heading outward in one of the arcs of stars leaving the Milky Way described above.

After the bulk of each galaxy has passed through the other, they will separate, but not permanently. Calculations reveal that as stars fly past each other, they create a gravitational drag on each other that effectively slows them down. Recall from Chapter 7 that as an especially massive star passes lower-mass ones, these are deflected behind the massive star and slowed down, thereby creating a clump of stars (Figure 7.3) that then pulls back on the star that created the clump with enough force to slow it down. Called dynamical friction, this prevents the stars in colliding galaxies from permanently flying apart, as they would have done otherwise.

With most of their stars still bound to them, but flying around now in all directions, the galaxies finish crossing each other and begin drifting apart, but without enough energy to separate far. They will slowly circle back and collide again. This second interaction will be more sedate, in that much of the gas is already either forming new stars or being blown out of the system, and the stars are moving more slowly. This time they won't go far apart.

Galactic Nuclei and Interactions

Most galaxies are believed to have supermassive black holes containing between a few million and a few billion solar masses in their very centers, their nuclei. As we discussed in Chapter 7, our Milky Way has a relatively low-mass supermassive black hole of about 4 million solar masses in its nucleus. Andromeda's supermassive black hole weighs in at around 140 million solar masses. Although the galaxies are unlikely to be on a path that will cause their black holes to immediately collide, dynamical friction will eventually bring them into orbit around each other. The two black holes will then spiral together in one of the most spectacular events of the collision process.

The merging process for the black holes begins as the nuclear regions of the two galaxies settle into orbit around each other. Unlike the Earth orbiting the Sun for billions of years in essentially the same orbit, the black holes will immediately begin spiraling toward each other. This occurs as the result of an effect predicted by Einstein's theory of general relativity. It says that whenever any two objects are orbiting each other, they emit gravitational radiation. These are ripples in the fabric of spacetime that travel at the speed of light. As a result of this radiation, the two orbiting bodies lose angular momentum, which causes them to spiral toward each other.*

The effect of gravitational radiation on a pair of orbiting objects was first observed in 1974 by Russell Alan Hulse and Joseph Hooton Taylor, Jr. Hulse

---

* The gravitational radiation created by the Earth–Sun interaction or the Earth–Moon interaction is too small to noticeably affect their orbits.

and Taylor observed that two neutron stars (cores of exploded stars that originally had masses between nine and twenty-five times that of the Sun) in orbit around each other were spiraling toward each other at just the rate predicted by general relativity, earning them the 1993 Nobel Prize in Physics.

Returning to the supermassive black holes in our colliding galaxies, depending on their initial orbit, their merger could take millions of years or more. The final stages of the black hole–black hole interaction are still being explored with state-of-the-art computer simulations. One thing that is certain is that their merger will be accompanied by a huge burst of gravitational radiation. Astronomers are on the verge of being able to detect such events occurring in other galactic mergers.

Another thing we know is that the resulting combined black hole will be larger and more massive than either of the two from which it formed. Because the Milky Way's supermassive black hole is roughly thirty-five times less massive than Andromeda's black hole, the combined black hole will be only slight larger and more massive than that of Andromeda today.

One of the more intriguing results of general relativity concerning the gravitational radiation emitted by the merger of these black holes is that the combined black hole can get a push from this radiation that will send it out of the nucleus. Depending on how much of a kick it gets, the black hole will ramble around the inner regions of the galaxy, wreaking havoc on all the stars, gas, and dust in its path. Fortunately, I posit that the solar system never wanders into this area. Nevertheless, as we show shortly, this black hole can still have a deadly impact on the Earth.

For hundreds of millions of years the merging galaxies will interact, helping force the myriad stars and other components to form a single galaxy with stars in random orbits in all directions, rather than mostly circling in disks as they were in the original galaxies. Over the lives of ten million generations of humans on Earth, the Milky Way and Andromeda will morph into an elliptical galaxy I call Milwan. Gone will be the disks. Gone will be the spiral arms. Gone will be the dense regions of gas and dust from which new stars can form. Gone will be the band of light in Earth's night sky.

If this collision occurs as currently expected, then in four billion years Milwan will be one of the largest galaxies in the universe, a giant elliptical galaxy. The dark matter halo plays an important role in establishing and maintaining the overall elliptical shape and size of Milwan. Therefore, I set the visible matter in this new galaxy to be embedded in the combined dark matter distribution resulting from the merger, rather than off to one side of it.

Fate of the Earth

The solar system will drift outward after the initial collision between the disks of the two galaxies, as I have discussed. People will see two bright regions in the sky representing the central regions of both galaxies. Over eons they will drift apart, but eventually the two supermassive black holes and their entourages of stars will head back toward each other. Viewers will see them merge and, in a very complex choreography, the two black holes will settle into orbit around each other. From Earth, that will mean there will be one bright center of the combined galaxy, Milwan.

While the merger described earlier is taking place, the gravitational attraction of all the matter in the combined galaxies will cause the solar system to stop moving away from Milwan. Earth will go around the central region of the new galaxy, but much farther out from the center than we are from the center of the Milky Way today. Furthermore, we will be on a more elliptical orbit than we are now.

The merged supermassive black hole wandering in the central region of Milwan has a potentially lethal role to play in the future of the Earth. This object will collect stars and gases as it rattles around the central regions of the new galaxy. Indeed, even today we see stars in orbits that are amazingly short and tight around the relatively small supermassive black hole in the center of the Milky Way (see Figure 7.2). Milwan's combined supermassive black hole will gravitationally attract a lot more matter, pulling apart stars and collecting gases that will plummet inward.

The supermassive black hole is a hole in the universe into which these gases can, in principle, fall. But there is a rub. Despite having 144 million solar masses, Milwan's black hole will be only 528 million miles across. That may seem large, but it is only slightly larger than the diameter of Jupiter's orbit around the Sun, which on the cosmic scale of things is tiny. There is so much gas being pulled toward the black hole that it can't all fall in as soon as it arrives; incoming gas collides with gas already waiting its turn to cross the black hole's boundary.

The incoming gas therefore forms what is called an accretion disk around the black hole, analogous to the region of swirling water waiting to go down a drain in your bathtub. As these gases spiral closer and closer to the black hole they are compressed into smaller and smaller volumes of space around it (Figure 10.4a). In this process the gases heat to over 180,000°F. At these temperatures, the electrons orbiting atoms in these gases are stripped off, creating a plasma. As the plasma swirls inward, it creates a magnetic field that spirals outward perpendicular to the plane of the accretion disk (Figure 10.4b).

Combining the hot gas and the magnetic fields, we get the final player in this drama: increasing temperature is accompanied by increasing pressure. That pressure becomes so great that some of the gas is able to push out of the accretion disk before it enters the black hole. This outflow occurs where the resistance to the gas is lowest, namely perpendicular to the disk. The result is hot gas that squirts out and is guided by the magnetic fields into two jets pointing in opposite directions, as you can see in the figure. So

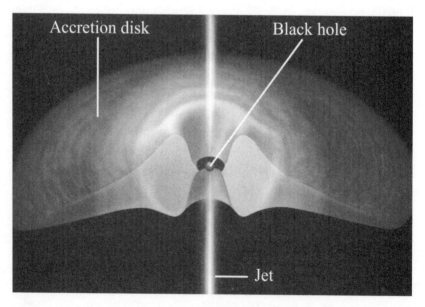

Figure 10.4a: Accretion disk around a black hole heats the in-falling gases and creates a pair of jets. CREDIT: W.H. FREEMAN & CO.

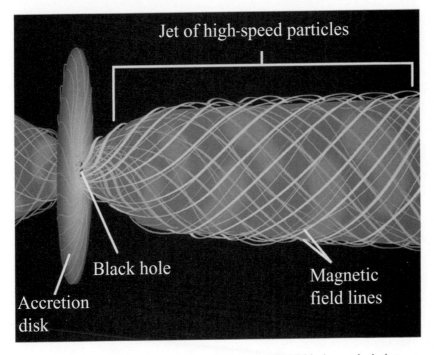

Fig 10.4b: *The hot gases in the accretion disk create magnetic fields that guide the hot, out-flowing gases.* CREDIT: W.H. FREEMAN & CO.

hot and intense are these jets that if the Earth were to be struck by one of them as it orbits Milwan, the high-speed particles in the jet along with the x-rays and other lethal radiation they emit would probably destroy all life on Earth.

I choose to put Earth in an orbit that does not ever pass through either jet.

Epilogue

As of the time I wrote this book, astronomers had not yet discovered life on any world other than our own. That may soon change as our technology explores Mars and as it reveals more and more planets orbiting other stars in our galaxy. But what if all that powerful technology is unable to detect extraterrestrial life? Does that mean we are alone in the universe? Does it mean that our planet is uniquely suited to support life? Does it mean that God created the Earth for life, and for humans in particular?

When I wrote *What If the Moon Didn't Exist?* back in the early 1990s, it was my intention to show how different astronomical conditions affect Earthlike worlds. I would be the first to admit that many of those worlds, along with many in this book, are not as appealing as habitats for humans as is the Earth. Creationists grabbed onto that first book and its inhospitable worlds as added evidence that God must have made Earth as a special place for life. If Earth were the only planet in the universe, then I would be inclined to agree with them. But it is absolutely clear that there are many planets, and we have already begun finding some, such as Gliese 581 d, in the Habitable Zones of other stars.

There are hundreds of billions of stars in typical galaxies and billions of galaxies in the universe. Based on the ongoing discoveries of planets in our cosmologically tiny neighborhood of the Milky Way, it is likely that there are at least millions of billions* of planets throughout the universe. There are billions of stars suitable for supporting life on habitable planets and I believe it is likely that millions of those worlds are in appropriate orbits around those (Sunlike) stars.

I therefore believe that we are not alone, nor are we necessarily living on the most ideal planet for life. Indeed, it isn't even clear what "ideal" means in this context. I think it is best to view evolution as making life more ideal for its respective planet. It is a tribute to the variety of atoms and their properties that life can form, change, and adapt as it has on Earth. I believe that the same is true for life on millions of other worlds.

Because science provides us with understanding, albeit incomplete, of every step in the star- and planet-forming processes, there is no reason to believe that divine intervention was necessary in creating the habitable Earth or any other world. At the same time, biologists are homing in on all the steps necessary for the formation and evolution of life. They haven't gotten as far, as fast, as astrophysicists in their understanding because life is far more complex than stars or planets. Nevertheless, it appears to me that the process of forming and evolving life will also be shown to follow natural laws without the need for divine intervention. If that is so, two profound questions come to mind: how did the universe begin and why is there life in it?

The question of how the universe began is still unanswered by science. That doesn't mean such an explanation can never be provided. Indeed, there are realms of mathematical research such as superstring theory that can meaningfully explore creation. Whether that work will lead to a testable, disprovable (i.e., scientific) explanation remains to be seen.

My perception for the second question is that life exists because it can. The same applies to self-aware creatures, such as ourselves. If that is true, it implies that life is what we make of it. Make the most of it. Asking "What if?" questions can help.

---

* That is, at least 10^{15}.

Appendix

TWO MOONS FORMED IN CO-CREATION WITH THEIR PLANET

Moons can form simultaneously with their planet. Stars and their host of orbiting bodies form when a tiny, slowly whirling fragment of a gigantic cloud of interstellar gas and dust begins to contract under the influence of the fragment's own gravitational attraction. The star forms in the center and as the remaining dust and gas are pulled inward, that debris begins to orbit the young star faster and faster, eventually forming a disk around it. The planets and other orbiting material condense in this disk.

This process of forming a disk with a large body (the star) at its center and smaller ones (the planets) orbiting around it apparently also occurs on smaller scales inside the disk when large planets, such as Jupiter, form and create miniature disks orbiting around themselves. The gas and dust they pull into orbit can form substantial moons, such as Jupiter's four largest moons, Io, Europa, Ganymede, and Callisto. The same may have also happened with Saturn and its largest moons.

Let's consider the evolution of two moons similar to our Moon formed in orbit in the same plane around an Earthlike planet. The planet is identical to the Earth today. The closer moon is at the same location as our Moon was originally, and the farther moon is twice as far away. We don't know the precise distance of our Moon when it formed, but plausibly it was ten times closer than it is today.

Both of the moons create tides on their planet, just as the Moon creates tides on the Earth. Because the planet is rotating (spinning) much faster than the moons orbit, the high tide closest to the nearest moon is pulled ahead of the moon by the planet's rotation (see Figure 1.1d). The water in this tide has a gravitational attraction to that moon pulling it forward in its orbit. This, in turn, gives the moon extra energy, which causes it to spiral outward.

Likewise, the more distant moon is pulled forward and outward by the tides, but more slowly than the closer moon. As a result, the inner moon spirals out to collide with the outer moon. Depending on the speed of their impact, it is likely that most of the two moons will unite into one larger moon. This will happen long before life could evolve on the planet, which is why this scenario isn't used in Chapter 1.

Why then haven't Jupiter's four large Galilean moons, Io, Europa, Ganymede, and Callisto, which all formed in the same plane as each other and with masses similar to the mass of our Moon, not all merged to become one moon after 4.6 billion years (the present life of the solar system)? The answer lies in the concept of resonant orbits.

In resonant orbits, two moons are being pulled by the planet's tides and by the gravitational pull from each other so that they gain energy and spiral toward each other, only to then lose the same amount of energy and spiral apart hours later, after which this cycle repeats itself. In a perfectly resonant system, the moons would be locked in those orbits indefinitely. If the planet is much, much, much more massive than the moons, like Jupiter and its moons, then it can hold multiple moons in resonance essentially forever.

However, calculations and computer simulations reveal that resonant orbits in a system like the planet and its two moons we are considering in this book are unstable. Slowly, inexorably, the moons are pulled free of the resonances and the inner one continues its spiral toward the outer one.

TWO MOONS FISSIONING OFF DIMAAN

Another possibility for creating two moons is that the planet forms from gas and dust spinning so rapidly that some of that material is flung off, like people standing on the edge of a merry-go-round spinning so fast that they can't hold on. We can quickly dispatch this idea. The problem with it is that the energy required to fling off enough mass to create even one moon in orbit where our Moon started its life is far more than is likely to have been available. Flinging off the material to form two moons requires much, much more energy (because the second moon would have had to go out farther than the first) and hence it is more than somewhat unlikely to occur.

Although the idea of a single moon forming this way was championed by Sir George Howard Darwin (1845–1912), son of Emma and (naturalist) Charles Darwin, as the way the Earth's Moon formed, he was wrong and there is no evidence that fissioning to form one or more moons ever happened in our solar system or is likely to occur elsewhere.

Selected Bibliography

Broecker, W. S., *How to Build a Habitable Planet* (Palisades, N.Y.: Eldigio Press, 1985).

Canup, R. M. and H. F. Levison, "Evolution of a Terrestrial Multiple-Moon System," *The Astronomical Journal* 117 (January 1999): 603–620.

Cattermole, P., *Planetary Volcanism* (New York: Halsted Press, 1989).

Comins, N. F., *Discovering the Universe,* 9th edition (New York: W. H. Freeman: 2008).

————, *What If the Moon Didn't Exist? Voyages to Earths That Might Have Been* (New York: HarperCollins, 1993).

Dole, S. H. and I. Asimov, *Planets for Man* (New York: Random House, 1964).

Ellis, J. and D. N. Schramm, "Could a Nearby Supernova Explosion Have Caused a Mass Extinction?" *Proceedings of the National Academy of Science* 92 (January 1995): 235–238.

Lewis, J. S., *Physics and Chemistry of the Solar System,* revised edition (San Diego, Ca.: Academic Press, 1997).

Lineweaver, C. H., Y. Fenner, and B. K. Gibson, "The Galactic Habitable Zone and the Age Distribution of Complex Life in the Milky Way," *Science* 303: (January 2, 2004): 59–62.

Littmann, M., K. Willcox, and F. Espenak, *Totality: Eclipses of the Sun,* 2nd edition (New York: Oxford University Press, 1999).

Miner, E. D. and R. R. Wessen, *Neptune: The Planet, Rings, and Satellites* (Chichester, UK: Praxis, 2002).

Murck, B. W., B. J. Skinner, and S. C. Porter, *Dangerous Earth: An Introduction to Geologic Hazards* (New York: John Wiley and Sons, 1997).

National Oceanographic and Atmospheric Administration, *Our Restless Tides,* www.co-ops .nos.noaa.gov/restles1.html (retrieved February 12, 2009).

Quintana, E. V. and J. J. Lissauer, "Terrestrial Planet Formation in Binary Star Systems," paper submitted to ExoPlanet task force May 23, 2007, http://arxiv.org/ftp/arxiv/ papers/0704/0704.0832.pdf arXiv:0705.3444v1.

Ribas, I., "Masses and Radii of Low-Mass Stars: Theory Versus Observations," *Astrophysics and Space Science* 304 (August 2006): 1–4.

Righter, K., "Not So Rare Earth? New Developments in Understanding the Origin of the Earth and Moon," *Chemie der Erde—Geochemistry* 67: 3 (October 2007), 179–252.

Sagan, C. and A. Druyan, *Comets* (New York: Random House, 1985).

Salo, H. and C. F. Yoder, "The Dynamics of Coorbital Satellite Systems," 1988, *Astronomy and Astrophysics* 205 (1988): 309–327.

Skinner, B. J. and S. C. Proter, *Physical Geology* (New York: John Wiley & Sons, 1987).

Voet, D. and J. G. Voet, *Biochemistry* (New York: John Wiley & Sons, 1990).

Zahnle, K., N. Arndt, C. Cockell, A. Halliday, E. Nisbet, F. Selsis, and N. H. Sleep, "Emergence of a Habitable Planet," *Space Sci Rev* 129 (2007): 35–78. DOI 10.1007/ s11214-007-9225-z.

Zeilik, M. and S. A. Gregory, *Introductory Astronomy and Astrophysics*, 4th edition (Fort Worth, Tex.: Saunders College Publishing, 1998).

Index

Page numbers in italics refer to illustrations.